The Geomorphology of French Landscapes

Adrian Harvey

The Geomorphology
of French Landscapes

 Springer

Adrian Harvey
School of Environmental Sciences (Formerly of the Department of Geography)
University of Liverpool
Liverpool, UK

ISBN 978-3-031-68489-0   ISBN 978-3-031-68490-6   (eBook)
https://doi.org/10.1007/978-3-031-68490-6

This Springer imprint is published by the registered company Springer Nature Switzerland AG
The registered company address is: Gewerbestrasse 11, 6330 Cham, Switzerland

If disposing of this product, please recycle the paper.

Preface

This book deals with landscapes. It has two underlying themes. Firstly, it deals with the science behind any explanation of the surface features of the earth that comprise "landscapes". That is the science of *Geomorphology*. This is done with particular reference to French landscapes. The science of Geomorphology sits somewhere between the academic disciplines of Physical Geography/Environmental Science and Geology. In the first part of the book I deal with some of the basic concepts of those disciplines insofar as they are relevant to understanding the landscapes of France. If readers feel that their own background in these fields is deficient I recommend two introductory books, both published by Dunedin Academic Press, which may help. These are *Introducing Geology* by Graham Park and *Introducing Geomorphology* by myself.

Secondly, this book is aimed primarily at a British readership. There are a number of books, both popular and science-based, that deal with British landscapes, but almost none (in English) that deal with the landscapes of our nearest neighbour country, just across the Channel, France (even in French there are not a lot!). The second part of the book deals specifically with the landscapes of France, region by region. Compared with the UK France is a large country, more than twice the size of the UK. Obviously, it would be unlikely for any author to know the whole of the country equally well. I know that within the text I have focused particularly on the regions that I know best or that hold more intrinsic geomorphological interest for me, but I have tried to give at least a minimum cover to all regions. A general point I would make relates to French place names, the consistency of their spelling (especially in the use of Saint/Sainte or their abbreviations St/Ste—with or without full stops or hyphens), plus other hyphenation and the use of accents. These may differ between IGN maps, Michelin maps, and local usage. Here I have tried to be consistent. As far as I could I have followed the usage on Michelin maps, plus I have tried to be consistent, using the St and Ste versions rather than the full versions of Saint/Sainte.

I suspect that there are two types of potential readership for this book, the academic and the general. The academic readership may include students or teachers visiting France professionally, who need an overview either of the geomorphology of France as a whole or of specific regions. This book is not a research tome as such but might point towards some research directions. For the general readership it seems to me that there is a gap in the field. I am thinking of the curious traveller who visits an area or a country and wants to learn about that area or country, so buys guidebooks. Guidebooks (regular bookshops seem to stock a lot of them) may mention "landscapes" in a shallow, trivial descriptive way, but offer no insight into landform processes, the geomorphology. The level of treatment is far below that given to historical or cultural material. At best such guidebooks may mention geology but even then they often "get it wrong" equating rocks with landscape (creating for example images of dinosaurs wandering around the Jura mountains—after all both are "Jurassic"). There seems to be no general appreciation that three phases are involved in the creation of landscapes—which may well be totally independent of one another and separated by huge spans of geological time. They are all simply described as "the geology". These three phases are (1) The formation of the rocks (in the French context involving timescales up to hundreds of millions of years). (2) Deformation of the rocks and their uplift by tectonic processes (the structure: in the French context involving timescales, particularly but not exclusively of up to around 70 million years). It would be correct to refer to both of these phases as "the geology". (3) The third phase is the erosion of these rocks and structures (and any associated deposition of sediments resulting from that erosion) to create the landforms we see today. This third phase involves processes, some of which are ongoing, which relate particularly to periods of only tens of thousands of years and almost wholly to periods of less than two million years. It is this third category that is most important in explaining the landscapes we see today. That is the basis of Geomorphology. Maybe this is a bee in my bonnet but to address that confusion was one of the factors motivating me to write this book.

Liverpool, UK   Adrian Harvey

Acknowledgements

I am immensely grateful to my wife, Karina, for her help and support, especially over the last few years, while I have been working on this book. I am also grateful to Sandra Mather, cartographer, formerly of the University of Liverpool, for her professional input and her patience. From the rather scruffy drafts I supplied her with she has produced a superb set of maps for this book. I thank my daughter, Fran Harvey, for her professional help in arranging slide scanning and Joshua Graham for his professional photographic input. I thank my son Michael for his invaluable help in accessing Landsat images. I also thank the Shutterstock FLEX trial programme for some images I was unable to otherwise source.

Contents

About the Author

Adrian Harvey has a background in Geomorphology and has had a lifelong interest in France. He graduated with a BSc in Geography with Geology from University College, London, in 1962 and then from the same institution with a PhD in Geomorphology and Hydrology in 1967. He was appointed to the Department of Geography, University of Liverpool, where he served for 40 years, before retiring in 2005 as Professor of Geomorphology. He has researched various aspects of slope and fluvial systems, particularly in the UK, the USA, and Spain. He has also served as President of the British Society for Geomorphology and acted as an Editor-in-Chief of the international journal *Geomorphology*. He has a long list of published academic papers as well as having published three previous books on various aspects of Geomorphology. His interest in France is more personal than academic and stems from his childhood. His father worked for British Railways and at the time BR had a cultural exchange programme with French Railways (SNCF), whereby the children of railway parents could exchange visits (with free travel). In this way Adrian Harvey spent half the summer school holidays, three weeks every year for five years, in France, mostly in the Alps and Provence, thanks to the Tournu family of Lyon. As an adult Adrian Harvey has maintained his French interests with family "Gite" holidays, numerous driving trips through France both en route to and from field research in Spain, and extensive trips purely for exploring France. Latterly he and his wife Karina maintained a small house as a second home in Burgundy. Needless to say, with this background he is appalled by BREXIT!

Part I
The Major Themes

Chapter 1
What Are Landscapes? The French Context

Introduction

What are landscapes? Usually we mean the visible features, both natural and human-influenced, of a tract of the earth's surface. The natural "physical" features would include mountains, hills, valleys, plains, rivers and coastal features (ie. the geomorphology), plus vegetation cover in so far as it is "natural" and not excessively managed by humans. The human landscape would include managed (semi-natural) vegetation including woodland and farmland, plus rural settlements (in the French context: villages). Our perception of landscape is conditioned by both its physical and human features.

The physical landscape (the geomorphology) depends in part on the underlying rocks and how they have been deformed to create geological structures (the geology), but the geomorphology is **NOT** the geology. The rocks and structures have developed over varying periods of up to several hundred million years. The geomorphology of the landscape that we see today has developed over a much shorter period of geological time, primarily through erosion of those rocks and structures, and to some extent through the deposition of sediments, the products of that erosion. The age of the resulting features of the physical landscape is perhaps several million years at most, and especially the detail only a few hundred thousand years, if that. In the popular literature, and even in some of the science-based literature, there is often confusion between the geology and the geomorphology. I do not deny the importance of the underlying geology (the rocks and structures) for **influencing** the development of the landscape. The geology is fundamental at the regional scale, accounting for example for the gross differences between say Brittany, the Alps, and Aquitaine. But within these regions the landforms (their geomorphology) depend on the history of the erosion and related deposition over much more recent timescales than those related to the formation of the rocks, or of the development of the structures.

© The Author(s), under exclusive license to Springer Nature
Switzerland AG 2025
A. Harvey, *The Geomorphology of French Landscapes*,
https://doi.org/10.1007/978-3-031-68490-6_1

The human landscape has evolved over even much shorter timescales than has the physical landscape. Humans (*Homo sapiens*) have been around in France for the last 40,000 years or so, but only leaving much evidence of their impact over the last few thousand years. Our perception of French landscapes, as expressed by the popular literature and tourist publications, is heavy on cultural influences based on history, architecture and the arts. Emphasis is often given to the (mostly human-modified) woodland vegetation cover, also to farmland, especially to vines (obviously "human" landscapes). In popular publications descriptions of the physical landscape are rarely science-based; at best they are based on geology, but that is usually almost entirely limited to rock-type (age of rocks), perhaps structure, but it almost never deals specifically with geomorphology (ie. erosional or depositional development of the physical landscape itself).

The primary aim of this book is to deal with the physical landscapes of France. I divide the book into two sections. The first section (Chaps. 2–5) deals systematically with the overall geomorphology of France. The second section (Chaps. 6–15) has a region-by-region emphasis, with the regions defined by their geomorphology rather than by politics or regional administration (Fig. 1.1).

France is a relatively large European country, similar in size to Spain or Germany and about twice the size of Britain. It has an enormous diversity of regional landscapes (Fig. 1.2). Much of this diversity relates to the physical landscape (the geomorphology), but also involves a diversity of the cultural (human) landscape. To a large extent the human landscape reflects the constraints and potential created by the physical landscape. I return to these themes within Chap. 5 where the relations between the physical and human landscapes are explored. However, our interpretation of those relationships depends to some extent on our perception of the landscape itself.

Our Perception of French Landscapes

Our perceptions of French landscapes relate to both the physical and human aspects of the landscape. These perceptions have been very much influenced by modern "coffee-table" books and guidebooks, usually containing superb landscape photographs. The emphasis in such books is primarily on the (human) cultural landscape. A little information may be given on the basic geology, but rarely is there any information on the geomorphology. Older guidebooks are much the same, albeit with poorer quality illustrations and again with more focus on the cultural than the physical landscape. Prior to the photo-age our perceptions would have been dependant on landscape painting, and these would perhaps be limited to the "art world" available only to those attending art exhibitions. Nevertheless landscape art would have been (and continues to be) influential in fostering interest in French landscapes.

Within the history of French painting, landscape painting has had a varied importance. Prior to the late eighteenth to early nineteenth century, it had only a minor role, at best in the form of cameos within paintings of other subjects. There were a

Fig. 1.1 Map showing the French geomorphic regions as defined in this book. The regional labels reflect the regionally-based chapters in the second part of this book (numbers refer to Chaps. 6–15), and also serve as locational information for material dealt with in the thematic first part of the book (Chaps. 1–5)

few exceptions such as grandiose depictions of Alpine mountain scenes, inspired by the grandeur of the mountains themselves (e.g. Turpin de Crisse was inspired by the glaciers of Mont Blanc: see Fig. 1.3a. He painted the Mer de Glace in 1810). This trend was continued later, in the nineteenth Century (e.g. by Gustave Doré, painting La Cirque de Gavarnie in the Pyrenees in 1882: see Fig. 1.3b).

By the early nineteenth Century, landscape painting had become established as a major theme in English painting, following the work of J.M.W.Turner (1775–1851), and John Constable (1776–1833). Turner painted an enormous range of subjects, including some that were landscape-based, for example after travelling in Europe painting "Hannibal crossing the Alps" (1812). Constable on the other hand focussed on rural landscapes mostly in his native Suffolk. There were really no equivalents in

Fig. 1.2 The diversity of French landscapes. (**a**) Mountains: the Mont Blanc Range seen from the Col des Aravis, Haute Savoie. (**b**) Plateaux: the southwest margins of the Massif Central dissected by the incised Tarn Valley at Ambialet. Note the level skyline related to a pre-Pleistocene erosion surface cutting across the tectonically deformed Hercynian rocks on the margins of the Massif Central. (**c**) Meandering rivers: The Lot valley west of Cahors, Aquitaine

Fig. 1.3 Landscapes that have inspired artists. (**a**) Modern glaciers on the north flank of Mont Blanc: the Glacier d'Argentière: Mont Blanc glaciers inspired Turpin de Crisse in the early nineteenth Century. (**b**) A Pleistocene mountain-glacially eroded landscape: the Cirque de Gavarnie in the Pyrenees, with a spectacular modern waterfall. This scene inspired Gustave Doré, in the late nineteenth Century. (**c**) Classic Chalk sea cliffs at Étretat, Normandy, including a stack and natural arch (the elephant's trunk!): This scene inspired Gustave Courbet in the mid nineteenth Century, and later Claude Monet. (**d**) Provençal mountains inspired Paul Cézanne: Mont Sainte Victoire ("Cézanne's mountain") near Aix-en-Provence. Painting by Cézanne, 1887 (from Maurice Raynal, 1959: "Cézanne, Biographical and critical studies", Editions d'Art, Albert Skira, Paris)

France until about mid-century. Then two landscape painters stand out, both known for painting realistically: Corot (1796–1875), who painted rural landscapes, and Gustave Courbet (1819–1877), better known for his erotic nudes, who did paint some French landscapes including the well known Étretat cliffs (1870) on the Normandy coast (see Fig. 1.3c). Both painters, in their different ways, can be seen as forerunners of the impressionist movement that dominated French painting during the latter part of the nineteenth century.

Most of the well known impressionist painters did paint landscapes, as well as other subjects (portraits, nudes, groups of people, urban scenes etc.), but their main concern was to capture the effects of light and movement. Édouard Manet (1832–1883) painted a few landscapes (mostly seascapes and riverscapes). Camille Pissaro (1830–1903) painted French rural landscapes, but focused on the vegetation and general scenery rather than on the landforms themselves. Alfred Sisley (1839–1899), an English painter, but working mostly in France, also painted rural scenes, particularly riverscapes. Auguste Renoir (1841–1919), best known for his portraits and nudes, did paint some Mediterranean seascapes. The two impressionist painters most focussed on landscape painting were perhaps Claude Monet (1840–1926) and Paul Cézanne (1839–1906) (See Fig. 1.3d). Monet, in addition to multiple works depicting the Étretat cliffs (1885) (see Fig. 1.3c), painted numerous waterscapes, especially on the Seine near his home at Giverny. Cézanne, a native of Aix-en-Provence, one of the few impressionist painters to focus on landform morphology, inspired by local Provençal landscapes, painted numerous studies of these landscapes, particularly of his local Mont Ste. Victoire (Fig. 1.3d). One other impressionist/post-impressionist artist who could be added here is Vincent Van Gogh (1853–1890), actually a Dutchman but working in France, worked for the latter part of his life in Provence and painted a number of dramatic Provençal landscapes.

The impressionists developed a whole new focus within landscape painting, particularly paying attention to the effects of light, especially on water and vegetation, rather than depicting the landforms themselves. They made an enormous contribution to the development of modern art, and paved the way for many new approaches that have characterised twentieth Century art (e.g. expressionism, cubism, and even abstract art). However, to my (limited) knowledge there is not a lot of modern (late twentieth century) French art that focuses on landscape, let alone on landforms. I am not aware of any French equivalent of the American, Georgia O'Keefe.

Enough digression into the cultural aspects of landscapes. The purpose of this book is to deal with the physical landscape itself, and how it evolved over the last few million years. The next three chapters (Chaps. 2–4) deal with basic principles of geology and geomorphology, especially in relation to the French situation. These chapters lead into a consideration of the relations between the physical and cultural landscapes of France (Chap. 5). The second part of the book is a regional treatment of the geomorphology of France.

Chapter 2
The Fundamental Geological Context for French Geomorphology

Background Concepts

This book deals with the landscapes of France, particularly with the landforms (the "Geomorphology"). In other words, it deals with what has governed the shape and relief of the landscape. Many guidebooks make the mistake of equating landscape-forming processes with the processes that formed the rocks of which the landscape is made. The most absurd example I have come across uses the phrase "this 500 million-year old waterfall" to describe a waterfall (not in France) that can only have been present for two million years at most, albeit over rocks that had formed in a totally different environment 500 million years previously. As another example, imagine the landscapes of the Jura Mountains—limestone ridges, cliffs in limestones, lakes and rivers within the valleys. These features were formed in relation to the glacial conditions that prevailed maybe most recently around 20,000 years ago, rather than the conditions maybe 150 million years ago under which the rocks themselves were formed in tropical/sub-tropical seas. Many guidebooks do not make this distinction, and refer to Jurassic landscapes, some even hinting that dinosaurs lived in these mountains! It would be analogous to books on medieval architecture emphasising rock quarrying or brick-making rather than building styles!

Having said that, we do need some understanding of the rocks of which the landforms are made and particularly of how these rocks may have been deformed by geological structures (folded, faulted) to create the "bones" of the present landscape. However, neither rocks nor structures are the landscape itself. The landscape has been created by erosion of the rocks which themselves may have been deformed by the structures. That erosion has resulted in creation of the erosional landscape. Some of the products of that erosion may have been deposited as sediments to form a depositional landscape. Quite clearly the age of the landscape relates to the erosional/depositional processes, rather than to the age of the rocks themselves, or to the age of their structural deformation.

© The Author(s), under exclusive license to Springer Nature Switzerland AG 2025
A. Harvey, *The Geomorphology of French Landscapes*,
https://doi.org/10.1007/978-3-031-68490-6_2

To fully understand the landscape, we do need to understand something of the origins of the rocks themselves, and of their structural deformation. This particularly relates to their elevation, because elevation is what conditions the overall relief of the landscape on which the erosional processes, which produce the topography, operate.

The contents of this chapter therefore, relate to the nature of the rocks themselves, and to their structural deformation, within the context of the geological timescale. I also consider briefly the plate-tectonics concept, the framework for understanding long-term continental evolution. I consider especially its application to the French context. Plate tectonics creates stresses within the earth's crust that result in deformation of the rocks. At a regional scale this results in lateral movement, uplift or depression, and at a local scale in compression or extension which results in the folding or fracturing (faulting) of the rocks. Finally I consider the later part of the geological timescale, the timescale during which the landscape itself has evolved, particularly through the relationships between geological and climatic processes.

If you are already familiar with these concepts, skip the rest of this Chapter and move directly on to Chap. 3, which deals with the geological sequence particularly in relation to France, or to Chap. 4, which deals with geomorphic processes and the landforms themselves.

The Rocks Themselves

There are three types of rocks depending on their mode of formation.

Igneous Rocks are formed by crystallisation from molten "magma": (i) at depth within the earth's crust (plutonic rocks, forming major intrusive bodies of rock, e.g. major granite masses); (ii) at intermediate depths (hyperbyssal rocks, forming minor intrusions—sills and dykes); (iii) at or near the surface (volcanic rocks, forming lavas and volcanic ashes). Igneous rocks are classified according to their mineral content. If they are rich in "light" minerals, high in silica (including minerals such as quartz and feldspar), and poor in iron-rich and magnesium-rich minerals, they are termed "acid" rocks (for example: granite). If feldspars, a complex family of silicates, are abundant, but there is no free quartz they are termed "intermediate" rocks (for example: dolerite). If they are dominantly of "heavier", dark minerals such as augite and olivine (again silicates), but rich in iron and magnesium together with some feldspar, they are termed "basic" rocks (for example basalt). Interestingly there is a general relationship between mineral composition and mineral weathering, with most of those minerals characteristic of basic rocks being chemically less stable than those characteristic of acid rocks.

Sedimentary Rocks, as their name implies, are lithified sediments (the sediment grains themselves derived from weathering, erosion and deposition from earlier rocks, then cemented together). Hence conglomerates (rounded particles) and breccias (angular particles) are derived from gravels. Sandstone, siltstone and shale/claystone obviously are derived from sand, silt and clay respectively. One particular type of sedimentary rock, related to particular sedimentary environments, is limestone. This is formed primarily by chemical precipitation of calcium carbonate (initially derived from seawater) rather than by clastic processes. Limestone occurs in a number of forms, some closely related to living organisms (e.g. coral). There is also a category of organic sedimentary rock—coal, derived from peaty organic material.

Metamorphic Rocks are rocks whose fundamental properties, which may include chemical composition, have been radically altered by heat and/or pressure. The metamorphism may be low grade, for example shale becomes a slate, sandstone becomes a quartzite, or higher grade involving much more radical alteration, recrystalisation and the forming of new minerals. During recrystalisation igneous rocks may be subject to mineral realignment (for example foliated granite). Such higher grade metamorphism involves recrystallisation, for example to form schist (relatively fine grained), or gneiss (coarse grained). In both the latter cases the nature of the original rock may not be obvious, even whether it was an igneous or a sedimentary rock. Metamorphism may result simply by association with local 'cooking' near volcanic centres, or be on a regional basis associated with intense tectonic activity related to mountain building.

The Geological Timescale (Fig. 2.1)

The geological timescale was originally devised in relation to fossil life forms characteristic of particular rocks. The oldest recognised fossiliferous rocks were assigned to the Palaeozoic (early life). Rock groups with more advanced life forms were assigned to the Mesozoic (middle life). Those with life forms not dissimilar from today's life forms were assigned to the Cenozoic (recent life). These divisions were subdivided into periods usually named after the region in which the rocks had first been described. The oldest of these (assigned to the early or Lower Palaeozoic) was named the Cambrian (these rocks were first described in Wales).

All rocks older than the Cambrian rocks were assigned the label "Precambrian". It was originally thought that the Cambrian was "the dawn of life" on Earth, but it has since been demonstrated that life had existed during at least the later part of the Precambrian. The timescale has now been reliably dated by radiometric methods, particularly using the Uranium content of zircon crystals in igneous rocks. The base of the Cambrian is dated to $c540$ million years BP (before present). The oldest known rocks on Earth (from the Canadian Shield, in northern Canada) date from about 4200 million years BP. This makes the Precambrian at least 7 times longer in duration than the whole of geological time since the Cambrian. This is all pretty irrelevant though in relation to French geology. There are no intact ancient Precambrian rocks in France, only youngest Precambrian rocks (mostly in Brittany) that have been incorporated into much younger structures.

EON	ERA	PERIOD		DATE MA BP (approx)	FRENCH EVENTS
PHANEROZOIC	CENOZOIC (TERTIARY)	Quaternary	Holocene	0.01	
			Pleistocene	2.5	Glaciations
		Neogene	Pliocene	5.3	
			Miocene	23	Alpine Tectonics
		Palaeogene	Oligocene	34	
			Eocene	56	Pyrenean Tectonics - Atlantic rifting
			Palaeocene	66	
	MESOZOIC	Cretaceous		145	
		Jurassic		201	
		Triassic		252	
	PALAEOZOIC — Upper	Permian		299	Hercynian / Armorican Tectonics
		Carboniferous		359	
		Devonian		419	
	PALAEOZOIC — Lower	Silurian		444	(Caledonian Tectonics NOT affecting France)
		Ordovician		485	
		Cambrian		541	
PRECAMBRIAN — PROTEROZOIC				2500	
PRECAMBRIAN — ARCHEAN				4000+	
		(Age of the Earth)		(4570) ?	

Fig. 2.1 The geological timescale. Note that the Cenozoic Era, subdivided into Palaeogene and Neogene, was formerly known as the "Tertiary". I use the terms "Tertiary" or "mid-Tertiary" in this book when the precision implied by the terms above does not match the dating of the events or the rock/sediment/landform sequences

The Basics of Plate Tectonics

Global-Scale Plate Tectonics The interior structure of the Earth has been established by the study of seismic waves, generated either by earthquakes or artificially by detonations. The waves are recorded at a receiving station some distance away. Their velocity and other properties depend on the nature of the material through which they have passed. What their study reveals is that the Earth is composed of a series of spheres. The core comprises an inner core, under very high pressure, of solid iron and nickel, and an outer core of the same composition but molten. The core is surrounded by the mantle, which is composed of ultrabasic silicate rocks in a plastic condition. Finally, there is the crust. There are two types of crust, a lower "oceanic" crust composed of heavier basic rocks, and an upper "continental" crust composed dominantly of lighter acidic rocks. The continental crust is spatially incomplete, and forms the basis of the major continents. It essentially "floats" on the underlying oceanic crust.

It had long been realised that over geological time the continents had occupied different positions. For example, the plan views of the eastern and western sides of the Atlantic Ocean exhibit an almost jigsaw-like fit, suggesting that they had once been one. It became clear that in the past the continents had moved around, broken up, and collided together. This concept was referred to as "continental drift", but until the 1960s no mechanism was known that could account for the phenomenon.

Following seismic studies of the crust, the theory of "plate tectonics" was developed. Under that theory new oceanic crust is being formed volcanically at mid oceanic ridges by seafloor spreading, for example at the modern mid-Atlantic ridge currently widening the Atlantic Ocean. The location of such mid-oceanic ridges is controlled by currents in the upwelling limbs of convection cells within the mantle (Fig. 2.2). Towards the corresponding downward limbs of the mantle convection cells the crust is drawn laterally downwards, underthrusting the adjacent crust, in what is known as a subduction zone. If the two crustal portions are oceanic crust then the surface expressions of this phenomenon are ocean trenches and volcanic island arcs (Fig. 2.2), for example today in the south Pacific. If, on the other hand, the crust beyond the subduction zone is of continental crust, there will be compression of the continental margin by thrusting and folding of the rock cover to produce a mountain range (Fig. 2.2), for example the modern Rocky Mountains and Coast Ranges of western Canada. In the extreme case of the two slabs of continental crust being brought together by this mechanism, there would be a continent-to-continent collision. This would produce a huge, very high mountain range, for example the modern Himalayas. When the mantle currents "switch off" or move elsewhere, the greatly thickened relatively light continental crust of the mountain zone is no longer being drawn down by subduction, but it responds buoyantly, "isostatically" by epeirogenic uplift, resulting in the great elevation of the major mountain ranges.

Regional/Local-Scale Structures Stresses induced by plate-tectonic activity are expressed at the regional scale by uplift or depression which tilt and/or fracture the rock bodies involved.

This process creates regional-scale structures: folds and faults. U-shaped (or basinal) downfolds are known as synclines, upfolded zones as anticlines (Fig. 2.3). Faults may be of several types. Most have relatively steep fault planes, either as extensional (normal) faults, or compressional (reverse) faults. Regional-scale tectonic compression may also result in low-angle thrust faults, especially at the plate-tectonic scale, such as those fronting thrust-forward mountain ranges (e.g. the Jura

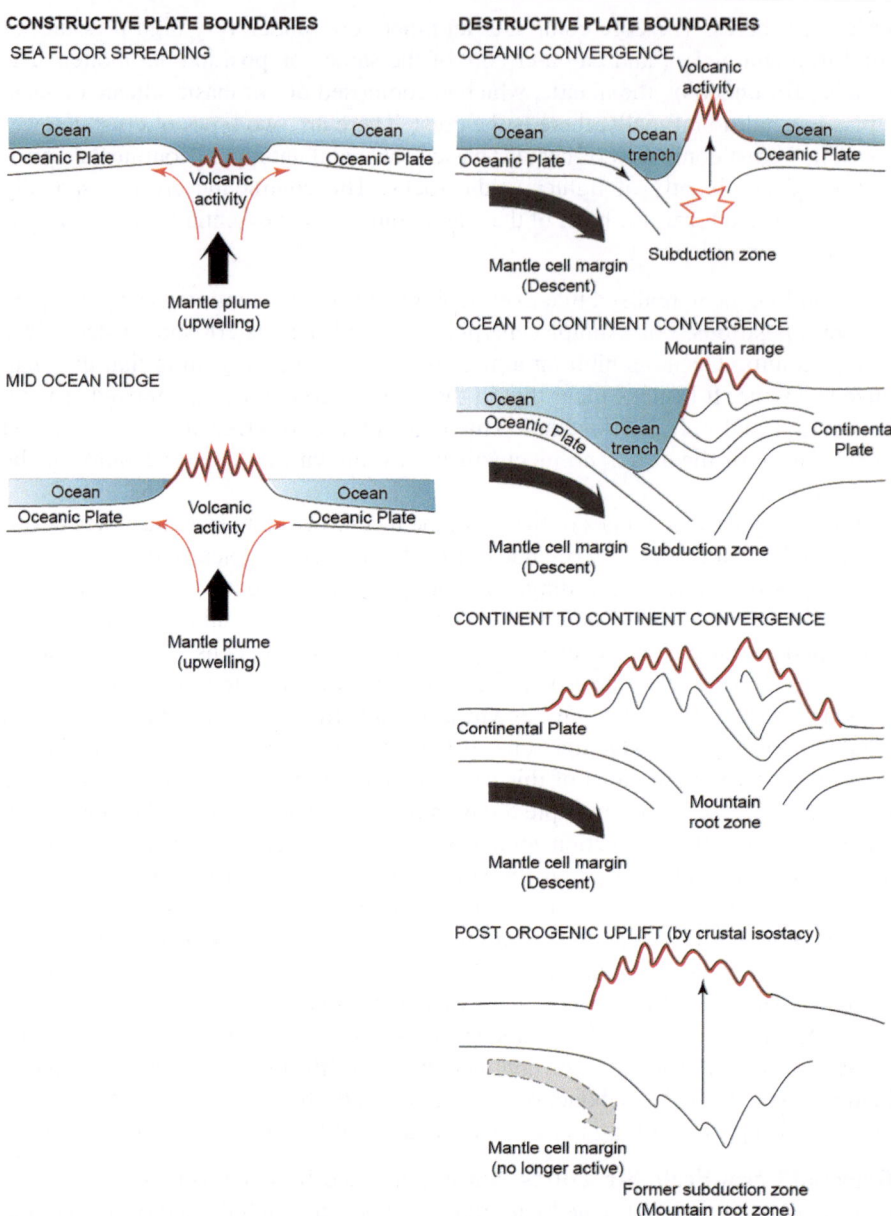

Fig. 2.2 The plate-tectonics model. Fundamental characteristics of constructive and destructive tectonic-plate boundaries

thrust front, and some of the internal thrusts within the Alps) (Fig. 2.3). Another important type of fault, associated especially with plate tectonics, is where the movement is primarily lateral rather than vertical (strike-slip faults). Within France

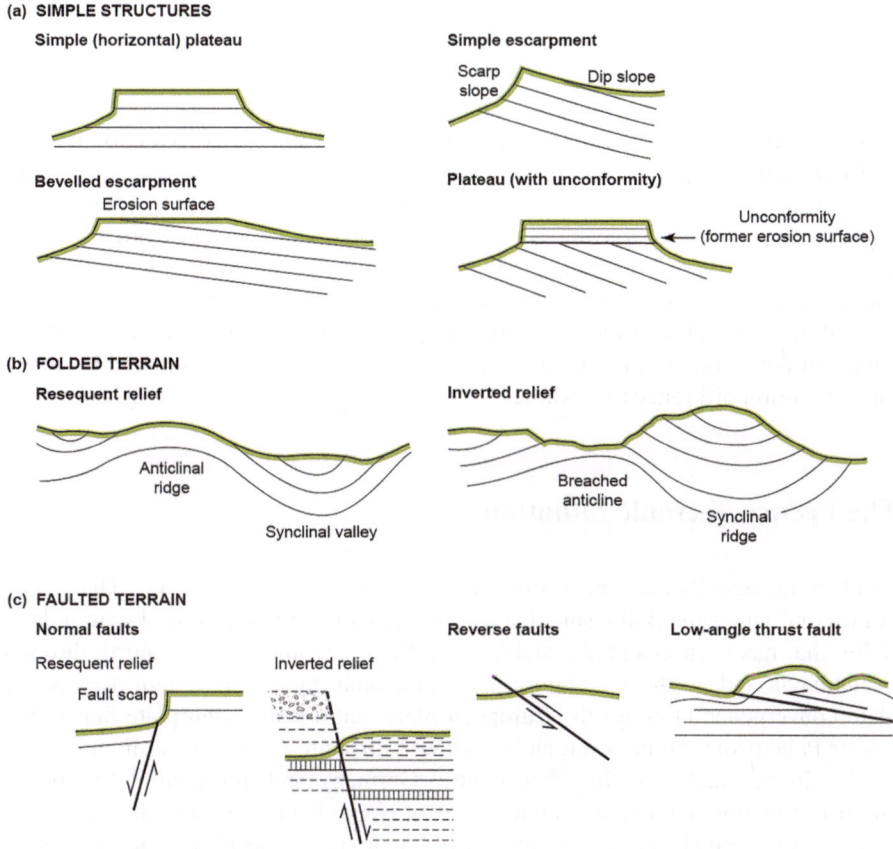

Fig. 2.3 Regional topographic expression of geological structures. Simple structures, Folded terrain, Faulted terrain

there are some such faults associated with Alpine tectonics, though often with considerable vertical movement too, but the clearest are perhaps those WNW-ESE Hercynian faults across southern Brittany (See Chap. 6).

On uplift above local base levels of erosion, the structures and the rocks associated with them are exposed to erosion at rates dependent at least in part on gradient and the available relief. The resulting landform assemblages reflect the relationships between the rocks themselves (their relative resistance to erosion), the structures, and the available relief. The resulting regional-scale landforms may reflect the structures directly (resequent relief) or inversely (inverted relief) (Fig. 2.3). Such relationships are exceptionally well developed in folded, or at least flexured, sedimentary terrain involving a range of rock resistances to erosion. They are well developed in the escarpment-dominated terrain of much of the Paris Basin, but also of the Jura, and of the Pyrenean and Alpine foothill zones.

So far, I have been stressing one major general rule within geomorphology. That rule is that neither the rocks nor the structures are actually the landforms. As a general rule, the vast majority of the landscape is the product of **erosion** of those rocks and structures. The age of the landscape is **NOT** that of the rocks, nor that of the structures, but relates to the timing of the erosion. One major exception to that general rule is the direct creation of landscapes by geological processes, particularly through volcanic activity. This is relevant in the French case in relation to the Neogene to Quaternary volcanic activity in Auvergne. I discuss those processes and features within Chap. 3 in relation to the timing of landform development in France, and in Chap. 8 specifically in relation to Auvergne.

With the general relationships between rocks, structures, and the gross scale of landform development in mind we can now consider the plate-tectonic and structural evolution of France as a whole.

The French Tectonic Situation

Past tectonic activity accounts for the gross geology of Western Europe. The "core" of the continent is the Baltic shield, a zone of ancient metamorphic rocks around the Baltic that has been essentially stable since the Precambrian. To its north through Scandinavia and northern Britain is the Caledonian mountain system, formed by plate convergence between the European plate and a Laurentian plate during the Lower Palaeozoic. To its south and southwest through western, central and north-eastern France and extending into central Germany and into central Europe, is another (now fragmented) mountain system of Upper Palaeozoic age, the Hercynian system (alternatively known as the Variscan or the Armorican system). For its French extent see Chap. 3. That mountain system was formed by collision between the then European/Laurentian plate and a palaeo-African plate.

The modern plate-tectonic situation affecting France has developed since the Mesozoic. I shall explore in more detail that evolution in the next chapter, but here I present the basic elements of the picture (Fig. 2.4). Sea-floor spreading has been opening the Atlantic Ocean since the Palaeocene (since about 60 million years ago). At the same time the European plate has been interacting with the African plate. During the Palaeogene these interactions caused the Iberian platelet to rotate from its earlier position in the Bay of Biscay, to impinge on the southern margin of the European plate and form the Pyrenean mountain zone. Later, during the Miocene (in the Neogene), the collision of the African plate with the European plate created the complexity of the Alpine zone affecting not only the French Alps but the whole of the Mediterranean region. Three other related aspects, that will be explored in more detail in the next chapter but should be mentioned here, are as follows. Firstly, the crustal thickening and buoyancy of the Alpine zone itself led to an increase in elevation. This effect has been accentuated by a further increase in buoyancy brought about by "erosional offloading". Secondly, the crustal stresses and strains that resulted from mountain building produced deep seated fracturing (faulting) in areas

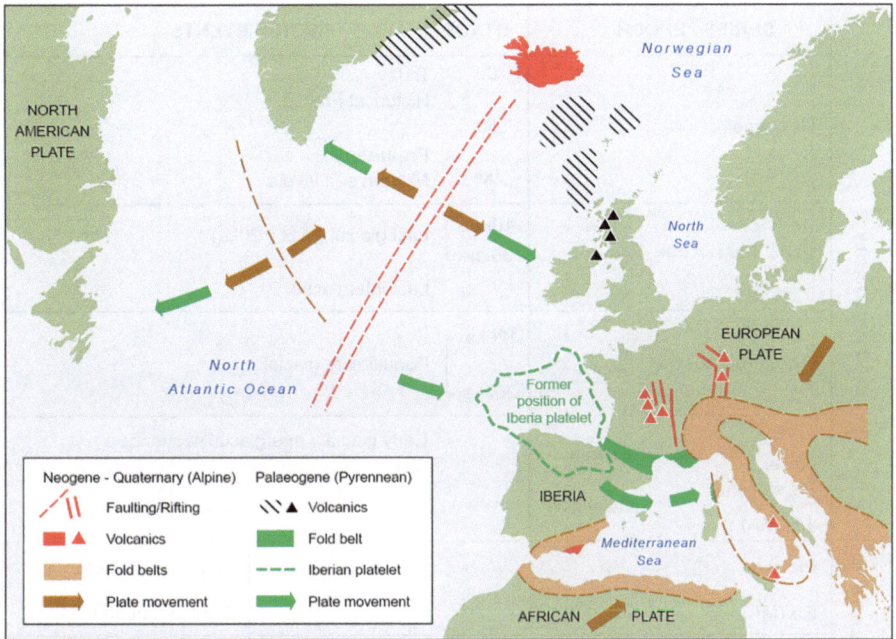

Fig. 2.4 Map illustrating the Cenozoic plate-tectonic evolution of Western Europe

some distance from the mountain zone. Thirdly, the effects of a major period of mountain building were expressed by less intense deformation (uplift and folding) away from the mountain zone itself.

The Neogene to Quaternary Timescale (Fig. 2.5)

The landforms we see today have resulted from the erosion of the rocks deformed by the geological structures, and the deposition of the materials/sediments derived from that erosion. The effectiveness of these erosion processes depends on interactions between internal (geological) processes and external (climatically-related) processes. Climatic processes themselves ultimately relate to the transfer of energy (heat) from tropical towards polar latitudes, partly by atmospheric processes and partly by ocean currents. Over much of the Mesozoic this mechanism was effective in maintaining relatively uniform global temperatures. However, during the Neogene the effectiveness of this mechanism of heat transfer was reduced as the result of the new locations of the continents. The position of Antarctica over the southern pole prevented ocean currents transferring heat into Antarctic latitudes. This resulted in a cold ice-cap dominated Antarctic continent. Furthermore, the almost land-locked nature of the Arctic Ocean prevented the penetration of warm ocean currents into Arctic latitudes, resulting in the formation of Arctic sea ice and the Greenland

	SERIES / EPOCH		STAGE	FRENCH EVENTS
Quaternary	Holocene		0 Today Historical France 2ka Pre-history 7ka Modern sea levels	
	Upper Pleistocene		10ka ⎤ 30ka ⎦ Last glacial (Max c.20ka) Last interglacial	
	Middle Pleistocene		125ka ⎤ ⎬ Penultimate glacial 200ka ⎦	
	Lower Pleistocene 1.8 (MA)	Sicilian Calabrian	Early glacial / Interglacial sequence Incision	
Neogene	Pliocene 5.3 (MA)	Zanclian		Erosion surfaces Mediterranean refloods
	7.1 (MA)	Messinian		Salinity crisis Mediterranean desicates
	Miocene 23 (MA)			Culmination of Alpine Tectonics

Fig. 2.5 The Neogene to Quaternary timescale. This relates to the period within which the French landscape has developed

continental ice sheet. Overall the result was a much greater temperature difference between polar and tropical latitudes than had previously been the case.

Another important climatic factor relates to orbital variations affecting the actual amount of solar radiation received by the Earth. These variations operate over roughly 100,000 year cycles, with a secondary oscillation over roughly half that timescale. The result during the Quaternary, given the reduced effectiveness of poleward heat transfer, has been climatic cycles alternating between "glacial" and "interglacial" conditions over these timescales (Fig. 2.5). Under "glacial" conditions large continental ice sheets formed over northern Europe and northern North America (in addition to the more-or-less permanent ice sheets over Antarctica and Greenland). Apart from a few small high altitude glaciers, the European and North American ice sheets melted during the (shorter) "interglacial" periods. The present interglacial has lasted for the last 10,000 years or so. [Incidentally, this has nothing to do with modern global warming—brought about by the changed atmospheric heat budget related to the atmospheric carbon content that has resulted particularly from the human use of fossil fuels.]

During Pleistocene glaciations France was not affected directly by continental-scale ice sheets, but there were smaller ice caps over the main mountain regions, particularly the Alps and the Pyrenees, but there were a few much smaller ice patches elsewhere (in the Vosges, in the Massif Central, and in Corsica). Elsewhere in France the cold conditions resulted in permafrost, permanently frozen ground at depth, but with surface layers that melted during the summer. This combination radically altered the surface geomorphic processes (see Chap. 3). There were of course vegetation changes also affecting geomorphic processes (see also Chap. 3). During the last 10,000 years (the Holocene) climates have been temperate in the north of France and Mediterranean in the south. Vegetation has re-established itself, and geomorphic processes have changed accordingly, only to be radically affected now by human activity, related not only to modern global warming, but to a whole human history of resource use and land use.

Chapter 3
France in Its European Geological and Geomorphological Setting

The Precambrian and the Lower Palaeozoic

France has an enormous variety of geology (Fig. 3.1) both of rock types and structural units. There are however, two important types of structural unit that occur elsewhere in Europe that are absent from France. Each continent has a core zone composed of Precambrian metamorphic rocks and complex structures that has remained structurally stable throughout Phanerozoic time (i.e. since the end of the Precambrian), the so-called "shield" areas. The core of the European continent is the Baltic shield (occurring obviously around the Baltic). There are also small fragments of another palaeo-continental shield, of a former "Laurentian" continent, occurring in the far north-west of Scotland. In the mid-Palaeozoic (Silurian-Devonian) a "northern" continent (Laurentia) converged with a "southern" continent (Gondwana), a collision referred to as an orogenesis, to form a mountain system, the so-called Caledonian mountain system. Remnants today form the basis of the mountains of Scandinavia and Scotland.

Neither shield areas nor remnants of the Caledonian mountain system occur in France. However, fragments of late Precambrian and Lower Palaeozoic rocks, deformed by the Caledonian orogeny, have been incorporated within the later Hercynian structures (see below).

A. Harvey, *The Geomorphology of French Landscapes*,
https://doi.org/10.1007/978-3-031-68490-6_3

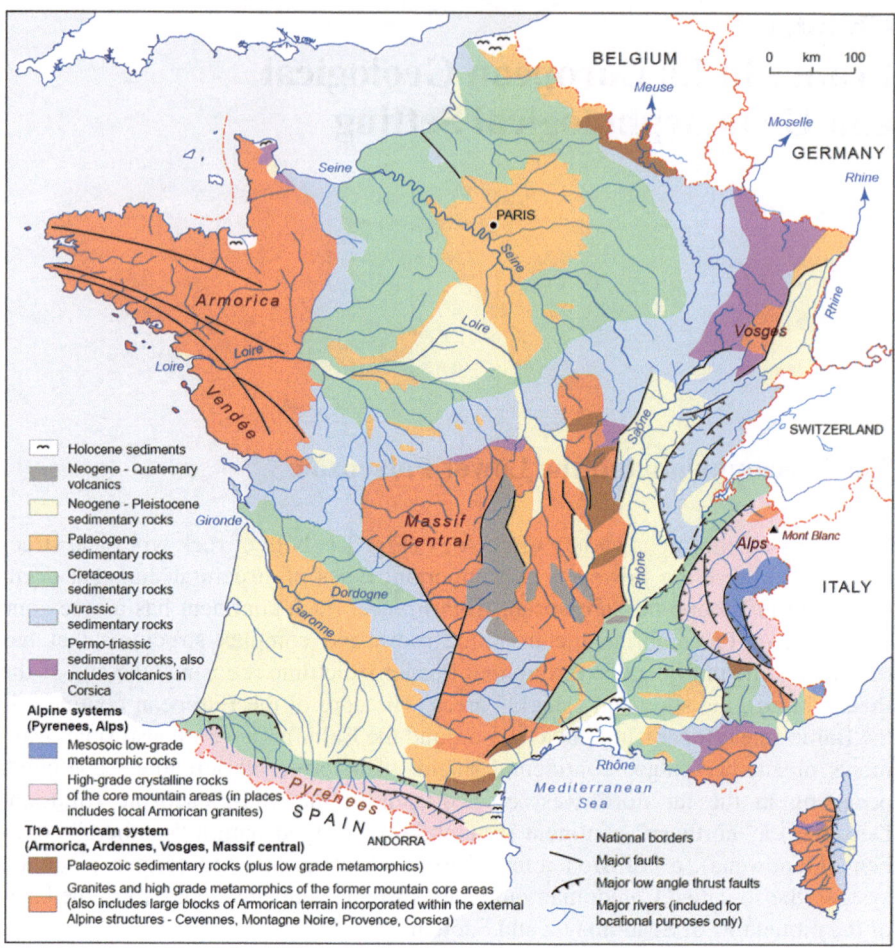

Fig. 3.1 Map showing the geology of France (Source: IGME: Institut de Géologie, Paris)

The Hercynian Mountain Systems

By the end of the Palaeozoic (the Permian) the southern continent (Gondwana) was again converging on the northern combined continent (Laurentia/Baltica). This created a complex mountain system, the Hercynian/Variscan system (the two terms are used interchangeably—also locally the term "Armorican" can be used). This system ran through what is now central Europe into France and Spain (which was then in the position of what is now the Bay of Biscay), clipping southwest England and southern Ireland and passing into America to form the Appalachians. (At the time, of course, the Atlantic Ocean did not exist.) Fragments of the earlier Caledonian system became incorporated within these Hercynian structures in Brittany and to a lesser extent in the Massif Central. Within France remnants of the Hercynian system today form the upland blocks of Brittany, the Ardennes, the Vosges, and the Massif Central. To the south, some rock units and Hercynian structures were later incorporated into the Alpine system of southern Europe.

Between the Hercynian blocks tectonic sag created the space for the later development of the Paris and Aquitaine basins (Fig. 3.1). Further southeast, space was created for the accumulation of Mesozoic and Palaeogene sediments that subsequently became incorporated into the Alpine structures.

The Mesozoic (Triassic, Jurassic, Cretaceous)

The Mesozoic was tectonically a relatively quiet era in Europe. Post-orogenic uplift of the new (Hercynian) mountain areas occurred, stimulating their erosion and the supply of sediment into the surrounding lower areas which underwent subsidence. In France these areas were particularly the Paris and Aquitaine basins, and in the southeast what later became the Alpine region (Fig. 3.1). During the Triassic, sedimentation was mostly as terrestrial sediment deposited in desert environments. With further subsidence of the basins marine sedimentation dominated the ensuing Jurassic and Cretaceous periods. For most of the time climates were sub-tropical resulting in carbonate sedimentation, producing the Jurassic and Cretaceous limestones that now dominate the geology of the Paris Basin, parts of southwestern France, the Jura and subalpine regions. In the Paris Basin (as in southeast England) these conditions culminated in deposition of the Upper Cretaceous Chalk formation.

The Palaeogene (Formerly Known as the "Early Tertiary")

Major changes took place at the Cretaceous/Palaeogene boundary (also known as the K/T boundary). These involved not only a major global faunal extinction (perhaps the result of a meteor impact in Central America?), but also major plate-tectonic changes that initiated our present geography (see Fig. 2.4). Seafloor spreading began to open up the Atlantic Ocean separating Europe from the Americas (with associated volcanicity in western Scotland). At the same time there was an anti-clockwise rotation of the Iberian platelet which had hitherto occupied a position within the (now) Bay of Biscay. It moved east then north. The result was the Palaeocene-Eocene compression, thrusting and uplift that formed the Pyrenean mountain structures. Related fold structures extended eastwards through the Cévennes and into southern Provence, the latter to be later incorporated within the "mid-Tertiary" Alpine system. Tectonic effects were felt far from the Pyrenees. There was movement along faults, and folding and flexuring of the Mesozoic sedimentary rocks of the Paris basin and in eastern France. During the Palaeogene relatively restricted marine sedimentation resumed in the Paris and Aquitaine basins (Fig. 3.1).

The "Mid Tertiary": The Alpine System (See Fig. 3.2)

By the Miocene, plate-tectonic convergence of Africa with southern Europe was occurring across the Mediterranean area creating the complex Alpine mountain system through Morocco, eastern Spain, the French-Swiss Alpine area, Italy, and east into the Balkans. Within France the dominantly Jurassic limestones of the Jura Mountains were folded and thrust northwestwards, probably over a mobile base in Triassic marls and evaporites (Fig. 3.2). Further south the dominantly Cretaceous limestones of the great arc of the Pre-Alps was thrust towards the Rhône valley, behind which the dominantly metamorphic rocks of the core area of the high French Alps were thrust upwards and forwards. Incorporated within these latter structures were zones of rocks which had previously formed part of the earlier Hercynian system. For example, in Corsica, the western Hercynian granites had earlier been plate-tectonically transported from the west creating the western Mediterranean basin. These were overthrust from the east by the "Alpine" rocks of eastern Corsica. The crustal thickening of all of these areas resulted in ongoing post-orogenic uplift, accentuated by the effects of erosional offloading, creating the great modern elevations of especially the high Alpine area.

Neotectonics and Volcanic Activity Away from the Alpine area itself uplift of the Massif Central and of the Vosges occurred. Major downwarping took place in basins marginal to these uplifted areas, forming the Rhine rift in Alsace, the Rhône-Saône valley system, also the two rift basins in the northern Massif Central, the Limagne (Allier) and upper Loire depressions. Major faulting occurred on the margins of these depressions, such as that bounding the eastern margin of the Massif Central, the Rhône-Saône fault system. Other fault systems included those bounding the Rhine rift valley and those bounding the two downfaulted troughs in the northern Massif Central.

Volcanic activity during the Mid-Late Tertiary and into the Pleistocene affected the German side of the Rhine Rift, and also occurred within the Massif Central in Auvergne and Cantal (Fig. 3.3a–c). There was further folding in the Paris basin and Eastern France, and the whole of France underwent regional uplift.

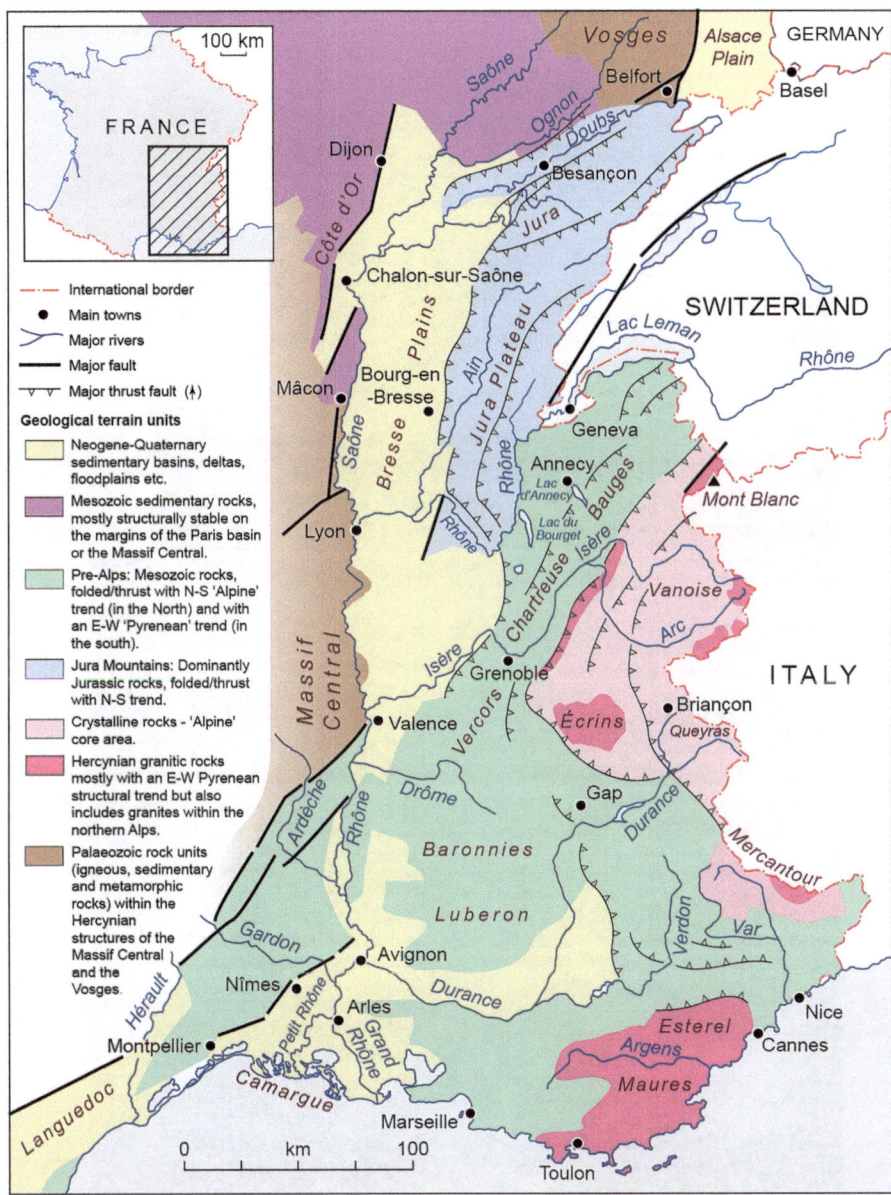

Fig. 3.2 Map showing the geology of the French Alps (Source: IGME: Institut de Géologie, Paris)

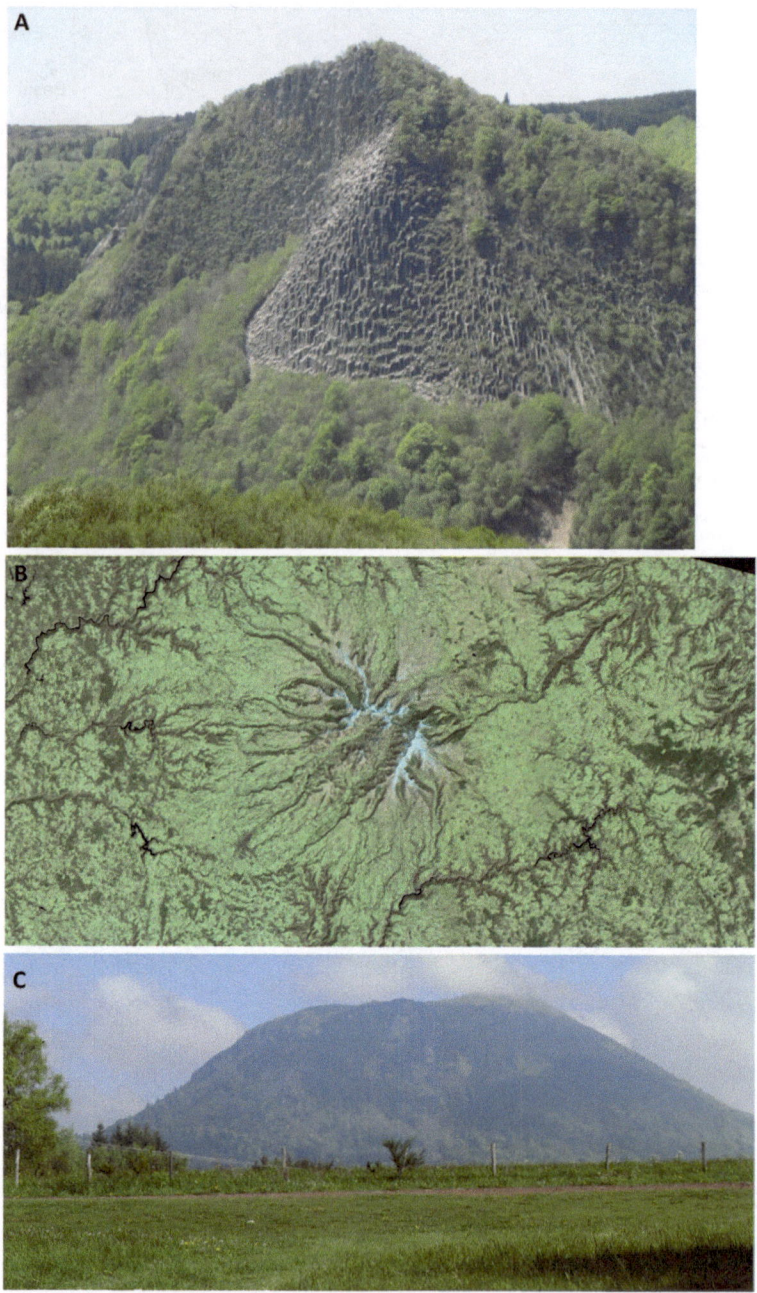

Fig. 3.3 Neogene to Quaternary landscapes (including the volcanics of Auvergne): see also Fig. 1.2b: Neogene erosion surfaces and Quaternary incision). This page: (**a**) Neogene-Quaternary columnar basalt: Roche Tuilière, near Mont-Dore, Auvergne. (**b**) Landsat image of the Miocene Cantal volcano, Puy Mary (E 2.45′ N 45.05′), southern Auvergne. (**c**) The iconic Puy de Dôme, a Late Neogene to Quaternary volcanic neck, Auvergne. This page: (**d**) Solutional pipes in the Chalk of Picardy: Filled with Neogene (?) "Terra Rossa" palaeosol

Fig. 3.3 (continued)

The Neogene (Formerly Known as the "Late Tertiary"): Relief Development

A curious event occurred at the end of the Miocene (during the Messinian). Tectonic activity had closed the outlet of the Mediterranean Sea to the Atlantic Ocean and the Mediterranean became a large saline lake. In its semi-desert climate it became subject to intense evaporation, precipitating salts, particularly gypsum, on the sea/lake floor and in marginal basins. Most of the information about this event comes from analyses of seismic studies and from sediment cores from the modern sea floor. However, in some areas (notably southeast Spain, Greece and southern Italy) there are gypsum exposures on land. These deposits were precipitated in marginal basins that have been subsequently tectonically elevated to be now exposed in terrestrial situations. I know of no similar sites in southern France. One aspect of the lowered sea level that is evident in southern France is the incision of the lower river courses, notably that of the Rhône, extending some way up the modern valley. The incised valley later became partially buried by fluvial and marine sediments, the forerunners of the modern Camargue delta.

A Major Phase of Erosion Much more widespread, were the effects of post-orogenic uplift, not only in the mountain areas, but also elsewhere. Between the Late Miocene and Mid Pliocene, shallow marine conditions in both the Paris and Aquitaine basins gave way to terrestrial environments. Slow uplift of both the Hercynian uplands and the Mesozoic rocks of the scarplands exposed both areas to erosion. The remnants of that major phase of erosion are preserved as gentle surfaces on the plateaux of the Hercynian uplands (Fig. 1.2b) and the scarplands. There has been debate as to the age of these "erosion surfaces". They truncate the underlying geological structures, and, where they occur within the Hercynian uplands, have been interpreted as perhaps ancient plains of unconformity (see

Fig. 2.3) since stripped of their cover rocks. The ages of the original erosion have variously been interpreted as sub-Jurassic, sub-Cretaceous, and sub-Eocene. Research on similar features in Britain however, suggests that they relate to Neogene erosion, rather than simply to the stripping of much older cover rock (see below: Chap. 4). The surfaces are often mantled by mature soils (Fig. 3.3d) which, in many places are buried by younger (Pleistocene) periglacial material.

The erosion surfaces are sometimes ill defined and in places appear to represent a sequence of successively lower surfaces, sloping away from the higher ground in the general directions of the major drainages. It was at that time (Late Miocene to Pliocene) that the major drainage basins became established. In southeast France these included the Rhône/Saône system; in western France, the Garonne/Dordogne and Charante systems. In northern France the major drainage basins included the Moselle/Meuse systems (ultimately part of the greater Rhine/North Sea drainage), and the Seine/Loire system (Fig. 3.4). At that time (Late Pliocene to Early Pleistocene) the Loire, downstream of a site to the east of Orléans, headed north following the valley of the modern diminutive Loing towards the Seine. Within the valley of that ancient river there are gravels containing clasts of Massif Central geology, that can only have been derived from the palaeo-Loire. The middle Loire was captured/diverted into its present westward course at some stage late in the Pliocene or early in the Pleistocene creating the space for the later formation of the swampy Solonge depression south of Orléans.

Neogene-Quaternary Volcanic Activity The other important Neogene process affecting part of France was the Miocene to Late Quaternary volcanic activity in the Auvergne. There were major effusions of lava, burying the previous landscape and creating basalt plateaux (See Fig. 3.3a–c), but most impressive was the creation of an enormous volcano in Cantal in southern Auvergne, during the Miocene (Fig. 3.3b). The main volcanic centre was around the modern peaks of Puy Mary and Plomb de Cantal. Volcanic activity continued in parts of the Auvergne, particularly the Le Puy and Puy de Dôme areas until the Late Pleistocene (Fig. 3.3c, see also Chap. 8).

The Quaternary

The Early Quaternary—Incision A major change took place at some stage in the early Pleistocene. Prior to that the main river systems appear to have been low gradient, non-incising rivers following the general topography expressed by the erosion surfaces. During the early Quaternary the main rivers, particularly in northern and western France, switched their behaviour to become incising rivers, creating deep valleys below the general low-gradient slopes above. The incised valleys often exhibit spectacular incised meanders (for locations see Fig. 3.5: see also Chap. 4).

Fig. 3.4 Map of the river systems of France

One ponders the possible causes for this dramatic switch to incision. There are several. First, Pleistocene sea levels were at times much lower than they had been previously. However, incision related to periodic base-level falls is unlikely given the low offshore gradients in what would have been the coastal areas. Furthermore, the actual incision primarily affects headwaters rather than downstream locations. Second, there may have been a tectonic cause, but I know of no direct evidence to suggest acceleration in post-orogenic uplift rate. Third, I suspect there was a climatic cause, with decreasing temperatures during periglacial conditions increasing runoff and therefore increasing stream power, triggering incision.

One aspect that might throw some light on the incision is the distribution and character of the incised meanders (Fig. 3.5). There are none in downfaulted basins (the Rhine rift valley, the Limagne/Allier and Upper Loire rifts in the northern

Fig. 3.5 Map of the river channel types in France

Massif Central, nor in the middle and lower Saône basin in Burgundy). There are none in the (formerly glaciated) northern Alps. Incised meanders do occur in the northern Jura (perhaps best developed outside the last glacial limits). They are extremely well developed within the Massif Central particularly on its western margins. They are also well developed in the Ardennes. They are quite well developed across the sedimentary rocks of the Paris and Aquitaine basins and in the eastern scarplands, but their distribution is sporadic (Fig. 3.5). On the lower reaches of the major rivers they are only well developed on the Seine.

One factor that seems to be important is available time. There has clearly been insufficient time since the last glaciation for the formation of incised meanders. The other possible cause (as suggested above for incision in general) may be ongoing tectonic uplift. It is interesting to compare their development and distribution with similar features in Britain, which tends to confirm some of the suggestions above. Incised meanders are much more widespread and much better developed in France than in Britain. In Britain they are more or less restricted to areas south of the last glacial limit. There are some north of that, but they are only poorly developed. The best developed in Britain are on the Hercynian rocks of Devon and Cornwall and on the lower Wye, with some less well developed on the Warwickshire/Worcestershire Avon and in the southern Cotswolds, all areas outside the last glacial limits.

Another important development during incision in specific areas was stream capture of which there are a number of examples in France (Fig. 3.4). That of the Loire has already been mentioned above. The best known examples are perhaps the capture of the previous upper Meuse by the Moselle near Toul in Lorraine (see Chap. 9), and the capture (or glacial/meltwater diversion) of the Doubs near Montbélliard from its previous northeasterly course towards the Rhine drainage into its modern westerly course towards the Saône (see Chap. 12). Also important is the capture of the upper Charente by the Vienne (see Chaps. 8, 10).

Pleistocene Glaciation The most significant Pleistocene development was climatic, with the onset of the Pleistocene sequence of global glacials and interglacials, over roughly 100,000-year timescales. We know much more about the last two glacial periods than about earlier glacials and interglacials. During glacials there was never any major continental ice sheet affecting France. In Europe there were only the British and the Scandinavian/Baltic ice sheets. However there was an Alpine/Jura ice cap, a much smaller ice cap over the Pyrenees, and several small glacial patches in the Massif Central and the Vosges (Fig. 3.6), probably also in Corsica (see Chap. 15).

Much of the land surface form in France owes its detail to the last global glacial phase. The direct effects of glaciation were intense erosion by cirque and valley glaciers in the high mountain source areas (Fig. 3.7a), with deposition dominant near the glacier limits, for example the spread of Alpine-derived glacial sediments in the Dombes area of the Saône valley north of Lyon (Fig. 3.7b), dating from the penultimate glacial, about 150–120 ka BP. The last glacial maximum was c 20,000 years ago, with the final switch to temperate postglacial climates c 10,000 years ago. Small remnant glaciers persist today especially in the higher Alpine areas, but also occur in small patches in the Pyrenees (Fig. 3.6), however even these are shrinking rapidly today as a result of global warming.

During the Pleistocene glacial phases, areas outside those directly glaciated were subject to periglacial conditions, with permanently frozen subsoils, locally creating patterned ground (Fig. 4.3d), but with seasonal melting of the surface layers. This seasonal melt produced slope instability, with unconsolidated sediments sludging downslope by a process known as solifluction (see Fig. 4.3e). Most of France was so affected. In addition, over the north European plain, including much of northern France, there was deposition of windblown silt (known as loëss) derived from the continental ice sheets further north and north-east. A further effect of glaciation was on rivers. Generally, during glacial phases sediment supply to river systems was high, even to those rivers draining only periglacial catchments. This caused most

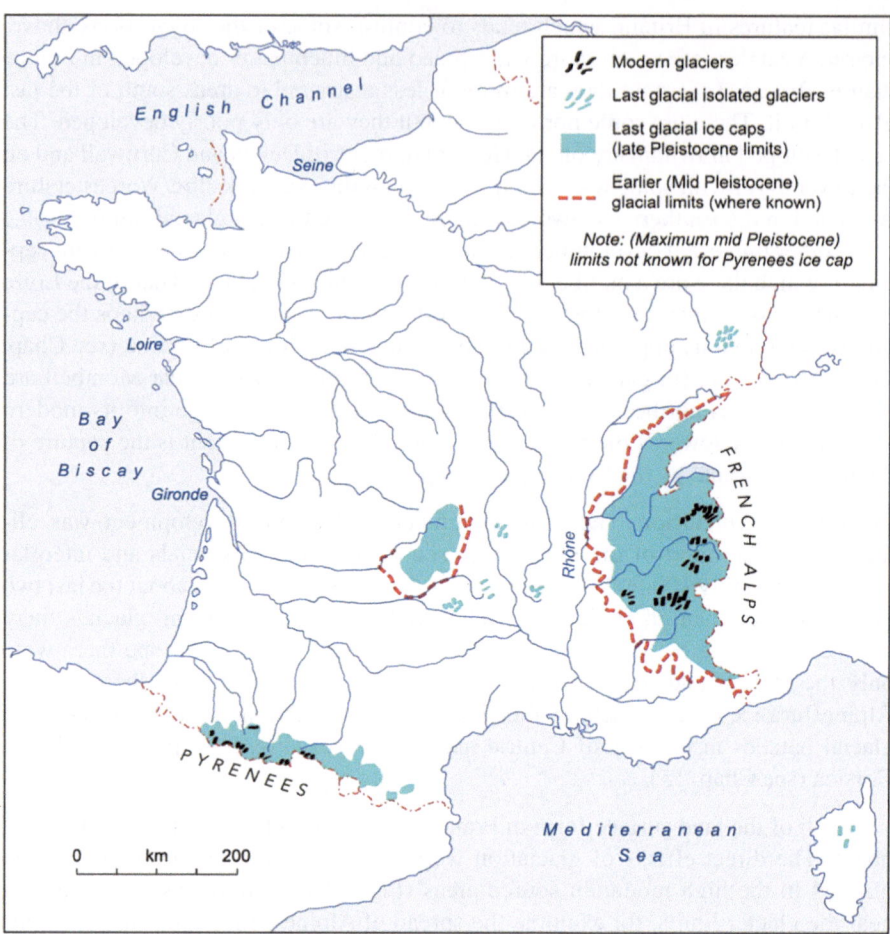

Fig. 3.6 Map of France showing the extent of Pleistocene glaciations. This map shows the maximum ice extent, where known, (relating to glaciation at about 150 ka BP), and the last glacial limits relating to the Last Glacial Maximum (at about 20 ka BP). Also shown are the locations of modern glaciers (Source: Mercier 2013)

rivers to aggrade during glacials, but incise during the intervening interglacials, creating river terrace landforms (see Fig. 3.7c, see also next section of this chapter, the Holocene; and Chap. 4: see also Pastré et al. 2003; Antoine et al. 2007).

There was another important effect of global glaciation. With significant quantities of the world's water stored on the continents within the ice sheets, there was a global lowering of sea level. During interglacials sea levels were high. For example during the last interglacial (somewhat over 100,000 years ago) sea levels were at or above present levels. Last-interglacial raised beaches or raised erosional rock platforms are preserved at a number of localities around the French coasts (Fig. 3.7d)

Fig. 3.7 Features related to Pleistocene glaciation. (For other views of modern glaciers and Pleistocene glacial topography see also Figs. 1.3a, b, 4.8a, b; and especially Chaps. 11 and 13). (**a**) A Large abandoned Pleistocene glacial cirque: Col d'Aubert, Néouvielle Massif, Pyrenees. (**b**) Morainic terrain, the Dombes—southeast of the Saône graben, Burgundy: Comprising grooved radial ridges with intervening linear swampy depressions. The morainic area was deposited during the penultimate Pleistocene glaciation (*c* 150 ka BP). Many of the intervening swampy depressions contain Holocene lakes which are home to many species of waterbirds. (**c**) A Pleistocene river terrace: The near-horizontal gravel surface above the Durance River, near Châteroux-les-Alps, north of Embrun, was the floodplain of a wide (extensively braided) river, related to the very high sediment supply during the Last Glaciation. Since then sediment loads have decreased and the river has incised, leaving the former floodplain as a terrace. (**d**) High sea levels during interglacials created what are now fossil raised beaches or raised rock platforms, as here in northern Corsica, a raised rock platform probably relating to high sea levels during the Last Interglacial (c 100,000 years ago)

During peak glacial periods, such as during the Last Glacial, the eastern English Channel did not exist nor did most of the North Sea (during the Last Glacial there was simply an ice-marginal lake further north). Britain was essentially a European peninsula to the north of France. Towards the end of the last global glaciation, about 15,000 years ago, sea levels were rising, reaching their present levels by about 7000 years ago. At some stage during the intervening period Britain became separated from the rest of Europe (a late Pleistocene BREXIT!).

An interesting overlap is that between the late Quaternary parts of the geological timescale (Pleistocene and Holocene) and the archaeological timescale relating to human evolution and pre-history. The early part of that timescale, the Palaeolithic (The Old Stone Age), broadly coincides with the Mid to Late Pleistocene. It was only during the Upper Palaeolithic (from about 40,000 BP), that modern humans (*Homo sapiens*) evolved, replacing earlier hominids, particularly the Neanderthals. The first clear evidence of *Homo sapiens* in France (skeletons and worked stone tools) dates from about that time. Later, from about 25,000 BP, during the Aurignacian phase, involving the so-called "Cro-Magnon man" (Early Modern Man) in the Dordogne valley, there was a cultural blossoming resulting in sculpture and cave paintings (in the famous caves at Pech Merle near Cahors, and Lascaux on the Dordogne). Later still, between about 20,000 and 17,000 BP, the Solutré culture (named after the site of its first description in southern Burgundy), became widespread. This was followed by the Magdelenian culture. The Mesolithic (Middle Stone Age) spanned the Pleistocene-Holocene boundary (c12,000–c7,000 BP) and was succeeded by the Neolithic (New Stone Age) during the Holocene (see below).

The Holocene Over the last 10,000 years, during the Holocene, two interacting sets of processes have been operating on the landscape: natural processes relating to stabilisation after and recovery from the last glaciation; and human-related processes, modifying many natural processes, and causing some overall destabilisation of the landscape.

Important amongst the natural processes was the post-glacial rise in sea level, which created the modern coastline. This coast has evolved over the last 7000 years or so into the modern patterns of cliffs, beaches and estuaries. The modern coast is largely a submerged coast (see Chap. 4). Only in areas of ongoing uplift is it emergent, and characterised by Holocene raised wave-cut platforms and Holocene raised beaches.

Inland in France the ice caps of the last glaciation melted, leaving a few remnant valley glaciers in the Alps and a few ice patches and small cirque glaciers in the Pyrenees (Fig. 3.6). No glacial ice remains in the Vosges nor in the Massif Central. The permafrost that was present during the last glaciation in most of France has melted, resulting in general slope stabilisation. The Pleistocene tundra-type vegetation was replaced by woodland, which together with warmer temperatures, resulted in a changed water balance and reduced runoff. This reduction, together with a great reduction in sediment loads being fed into streams and rivers, resulted in changes in river styles and patterns (see Chap. 4). Many rivers which in the late Pleistocene had had high sediment loads leading to aggradation of the valley floors, had their sediment supplies reduced in the Holocene. In most cases this resulted in incision into the former valley floors, which then became perched above river levels as river terraces (Fig. 3.7c). Braided channel patterns which hitherto had been widespread, became restricted to mountain streams and a few large rivers. Formerly (Pleistocene) aggrading valley floors were dissected as they became sediment starved, and now formed terraces above the newly developing (Holocene) valley floors.

Archaeologically, the mid Holocene (from about 7000 BP) saw the transition from the Middle Stone Age (the Mesolithic, see above) to the New Stone Age (the Neolithic). Not only did working in stone become much more sophisticated, but Neolithic people constructed megaliths (alignments of large stones, e.g. at Crozon in Brittany, not unlike Stonehenge in England). Later, from about 3000 BP, a variety of peoples brought metal working to France (the Bronze Age). Then around 2300

BP the Celts brought iron technology (the Iron Age). They were followed by Mediterranean peoples settling on the French Mediterranean coast, and eventually in 59–55 BC, by Julius Caesar and the Roman conquest, perhaps marking the "beginning" of modern History.

The human impact on geomorphic processes and on the landscape has been increasing over the last 2000 years or so, but especially over the last 150 years. From prehistoric times onwards the clearing of the mid-Holocene woodland for agricultural land would have affected the water balance, runoff and streamflow. In steep slopes in parts of the south of France, especially in Corsica, early agricultural activity involved the creation of agricultural terraces, especially for crop cultivation (?). One result would have been soil conservation. Elsewhere on other steep slopes forest removal or agricultural activity locally triggered erosion by overland flow, especially in the driest regions (Mediterranean mountain regions), creating gully and badland areas (see Chap. 4). Any change in streamflow and sediment supply to river systems would also have had implications for river channel types and patterns. In more recent years there has been some re-forestation, especially in some mountain areas, again affecting water-sediment relationships, therefore modifying river channel types and patterns.

More recently, basically since the industrial revolution, effectively over the last 150 years, there have been much more significant changes to geomorphic processes brought about by direct human intervention. Coastal protection works have modified coastal processes, as has saltmarsh reclamation. Mechanised agriculture, especially the ploughing of slopes, has modified sediment yield by increased hillslope erosion. Mining and industrial development have also had local impacts, but perhaps the most significant modifications have been to the river systems (Fig. 3.8). During the nineteenth Century canal development took place along and between most of the major rivers of France. In some cases this had only a minor impact on the rivers themselves. However, on those rivers where the channel became the main path for river/canal transport, the impact was greater (e.g. channel modification by straightening, bank protection, weirs affecting water levels and sediment transmission). Most important of all perhaps was dam construction (Fig. 3.8), not only in the mountain areas but also by the creation of artificial lakes in some lowland areas. Dams modify the flow regime with a tendency to reduce flood discharges, and most importantly of all they trap sediment. Hence, in many cases there has been a tendency for degradation of the sediment-starved river channel downstream of a dam. The Rhône system throughout its (French) length has been radically altered by dam construction and diversions for power generation. Of the larger rivers within France, only the Loire and the Saône have channels anything like their "natural" channels. A word of caution: not all the potential changes brought about by dam construction are slow and gradual; they may be catastrophic e.g. dam failure inducing catastrophic floods downstream, or landslide-induced overtopping and flooding as occurred on the Vajont dam in the Italian Alps in 1963, with disastrous downstream consequences.

Fig. 3.8 Map showing the extent and style of human impact on the river systems of France

All these aspects of human impact are effective at local and regional scales. In the long term perhaps they may be relatively minor, even when involving local human-induced environmental catastrophes, when compared with the potential colossal effects of modern global warming!

Chapter 4
Landforms and Geomorphic Processes (French Examples)

When considering the processes and the factors influencing the development of landforms two scales must be considered. First is the regional scale of gross landform development (see below) relating to such aspects as overall relief, mountain elevation, plateau topography, overall drainage patterns and valley configuration. The processes involved are tectonics-driven internal processes and their interaction with climate-driven external processes. Both sets of processes operate at regional scales over timescales of the Neogene onwards. In Chap. 3 I have dealt with the mechanisms involved. The second scale is the local scale of detailed landform development (see further below) through the processes of weathering, erosion, transport and deposition. These processes are largely climatically controlled, operating within a context provided by the gross landforms.

Long-term Gross Relief and Landform Development in France

Plateaux I have already dealt with the Alpine tectonic context that created the relief of the Alpine areas, and with the post-tectonic epeirogenic uplift that affected the elevations of those and other areas (Chap. 3). I have also dealt with the (probable) Neogene development of erosion surfaces especially across the ancient Hercynian massifs and the mostly Mesozoic-rock scarps and plateaux of northern and eastern France (Chap. 3, see also Fig. 2.3). However, there are several aspects of these features that need further consideration. First, these features are undoubtedly pre-Pleistocene in age. In many places in northern France they are overlain by periglacial loëss (windblown silt) of undoubtedly Pleistocene age. The loëss often buries older soils. Otherwise such soils are exposed at the surface. These are very mature red soils ("terra rossa") which take millennia of surface stability to form.

They probably formed under a climate rather warmer than that of today. On the Chalk plateaux such soils can also be seen occasionally piped down into solution features within the Chalk (see Fig. 3.3d). Within the Massif Central there are similar soils buried by Pliocene and Pleistocene lavas (see Chap. 8).

In both Britain and France the Neogene erosion surfaces formed a major research field in geomorphology up until about the 1960s. Since then, they have received very little research attention. There was debate about their possible origins. Are they uplifted peneplains, plains produced by "normal" slope and river processes under stable base-level conditions? Are they uplifted pediplains, similar features produced under semi-arid conditions? Or, are they uplifted etchplains, similar features affected by tropical or sub-tropical deep weathering? The soils would suggest the third possibility is perhaps the most likely. More recent work in the eastern Pyrenees, though detailed and confirming their Neogene age, is equivocal on the specific processes of their genesis (Calvet and Gunnell 2008). There is no clear answer to these questions. Whatever their origin, the processes maintaining them abruptly ceased early in the Pleistocene. Planation was replaced by deep dissection by the main river valleys. This was probably the result of climatic deterioration resulting in increased river flows. The incision was pulsed, often with earlier stages marked by within-valley benches, below which are sequences of river terraces (see below). Many of the rivers developed enormous incised meanders (see above, Chap. 3), within which in many cases the modern river is a diminutive remnant of its earlier self. The meanders of the modern channel are of much smaller dimensions than the valley meanders, a "misfit" situation, (see Fig. 3.5), the implications of which are discussed more fully below (see also Fig. 4.5d).

Incision The stages of incision were pulsed in response to Pleistocene climatic variations. Within the incised valleys are river terraces which preserve remnants of earlier valley floors and sediments (see below and Fig. 3.7c). Research over the last 15 years or so, especially by authors such as Pastré et al. (2003), Cordier et al. (2006), and Antoine et al. (2007) who have researched the incised valleys of northern and eastern France, has demonstrated up to eight terrace levels. Sediments of three of the terraces have been precisely dated to reveal a sequence tightly controlled by the climatic sequence of alternating glacial/periglacial phases (stadials) and interglacial (interstadial) phases (Antoine et al. 2007). During the interglacials/interstadials the rivers had small multiple, anastomosing channels carrying only fine sediments and the valley floors were swampy, a situation similar to that during the early Holocene. During the early part of the ensuing glacial/periglacial phase, discharges and stream power increased and the channels became erosional, initially migrating, then incising into the valley floor. In this way the original valley floor was transformed into a terrace. Later, at the peak of the glacial/periglacial stadial, sediment yields were high, particularly of gravel sediments, often forming braided channels. The rivers began to aggrade building a new (primarily gravel) valley floor. As climate warmed towards the ensuing interglacial/interstadial, discharges decreased, sediment yields also decreased and became dominated by fine sediments. This sequence produced a new swampy valley floor burying the gravel, similar to but set below that of the previous interglacial/interstadial.

Fluvial instability and change have continued during the Holocene. For the first part of the Holocene the rivers responded to changes in hydrology caused by continued climatic fluctuations and vegetation change. In the later part of the Holocene, and particularly in the last 150 years, human activity has had a dominant influence on fluvial processes and morphology (see below). For example, on a small stream In Normandy Lespez et al. (2008) have demonstrated a response to climatic fluctuations during the early and middle Holocene. They found variations in the style of sedimentation including some peat growth and some carbonate precipitation. From the iron-age onwards, soil erosion contributed much more silt, deposited as overbank sediment on the floodplain.

In contrast, several studies in the calcareous pre-Alps (Bravard 1989; Liébault and Piégay 2002; Liébault et al. 2002; Astrade et al. 2011) demonstrate more recent channel change. Following major gravel sedimentation during and after the cold conditions of the "Little Ice Age" (seventeenth–eighteenth centuries) human-induced environmental changes (some abandonment of agricultural land, but particularly re-afforestation) have reduced gravel sediment supply with the result that the channels narrowed, and in many places have scoured into bedrock.

Geomorphic Processes and Environments

When considering landforms and processes, the second scale is the detailed scale of surface form: the relief and patterns of the forms within the gross topography, and how these relate to geomorphic processes themselves (weathering, erosion, transport, deposition). Here, I consider these processes over timescales of the late Quaternary to currently-active, and at the detailed spatial scales of an individual valley or individual relief feature.

Weathering "Weathering" relates to the breakdown by mechanical, chemical or biological means of the parent rock and the subsequent alteration of the products of the primary weathering.

Mechanical Weathering A major primary mechanism of mechanical weathering is through what is known as the development of pressure-release jointing (see Fig. 15.4a). This is particularly important in the weathering of plutonic rocks (e.g. granite). These rocks are formed by chrystalisation at some depth below the earth's surface under conditions of higher ambient pressures than those at the surface. When tectonically uplifted and exposed at the surface by erosion this pressure is released resulting in cracking of the rocks in the form of near-surface surface-parallel fractures. These features are evident especially in tors, such as those on the upland surfaces in the granite terrain of interior Brittany (see Chap. 6). Such fractures allow the penetration of water into the body of the rock allowing the chemical weathering of the rock's constituent minerals (see below). Other good examples of pressure-release jointing that I know in France are in the Corsican granitic terrain in the western part of Corsica (Fig. 15.4a). Other granite areas would also be subject to the same processes. Pressure release as defined above would also include an element of "erosional offloading", whereby the weight of the overlying rocks is removed. Hence even sedimentary rocks, when uplifted, would be prone to "erosional offload-ing" resulting for example in the widening of bedding planes. Furthermore, it is not only tectonically-induced uplift or surface erosion that may induce offloading. Even glacial melt may act in this way.

There are several other mechanisms for mechanical weathering. The first, per-haps the most important, relates to freezing and thawing. Water that has penetrated into pre-existing spaces in the rock (e.g. bedding planes, joints, pore spaces) on freezing, expands. This causes internal stresses within the wetted rock to induce internal fractures. This process produces angular rock debris, the raw material of scree (Fig. 4.1a, see also Figs. 11.3c, 11.4a). Even today such processes are effec-tive at high elevations in the Alps and other mountain areas, but during the glacial phases of the Pleistocene they were especially effective on all exposed rock faces. There are three other simple mechanisms that may be effective in mechanical weathering. Wetting and drying of a rock may cause internal stresses sufficient to induce cracking, or in soft rocks, crumbling. Intense direct heating and cooling may also induce expansion and contraction of the rock surface sufficient to induce crack-ing. This is an important mechanism in desert environments, but probably not important in France either now or in the past (except perhaps in Mediterranean France during heatwave conditions?). The final mechanism is crack enlargement as a result of tree-root penetration into pre-existing cracks.

Chemical Weathering   The minerals within rocks or sediments may undergo chem-ical changes related to solution/precipitation, oxidation/reduction, hydration/dehy-dration, ion exchange, and biological/biochemical processes. The soluble products of chemical weathering tend to be removed in solution, either locally down-profile, or away from the site altogether. The insoluble products tend to be either the more chemically stable constituents of the original rock or the insoluble products result-ing from the chemical processes listed above. From sedimentary rocks these tend to be dominated by quartz (sand grains) and complex silicates (clay minerals), and perhaps iron oxides. Soluble minerals (especially $CaCO_3$) may be removed in solu-tion (see below). From igneous rocks the products will be quartz (if present in the original rock, e.g. granite), clay minerals and iron oxides (both derived from feld-spars and ferromagnesian minerals in the original rock).

Infiltrating rainwater may be slightly acidic therefore may dissolve soluble salts within the rock or sediment and leach them downwards. The reverse may take place under dry conditions. Dissolved salts may be drawn to the surface by capillarity, and precipitated at the surface as the moisture evaporates. Geomorphologically this is an important process in dry or seasonally dry environments such as in the Mediterranean climate of southern France. A common salt precipated in this way is Calcium

Fig. 4.1 Weathering and karst features. (**a** and **b**) Weathering features: (**a**) Frost shattering produces angular rock debris, the primary constituent of scree: Pyrenees. (**b**) Honeycomb weathering of granite in Corsica. Two stages are involved. First: the chemical breakdown of some of the constituent minerals to much weakened 'grus'. Second: the reprecipitation of the previously dissolved material under dry conditions to form the resistant shell—the case hardening. (**c** and **d**) Karst features: (**c**) Surface solutional features: a limestone pavement on Jurassic limestone in the Jura Mountains. (**d**) Subsurface features: caves produced by focused solution of Jurassic limestones: Azé caves, Mâconnais, southern Burgundy. Note the calcite precipitational features (e.g. tufa curtains) on the cave roof

Carbonate ($CaCO_3$). The resultant deposit, which may form a caprock over much weaker material, is known as calcrete. Its nature and maturity can be used as an aid to the relative dating of the surface at which it is deposited. A similar process, usually involving silica salts rather than $CaCO_3$ and producing silcrete rather than calcrete, may take place over weathered igneous rocks (e.g. granites). This is a process known as case-hardening. Further weathering and erosion of the weakened sub-surface weathered material may result in a surface characterised by "honeycomb weathering", a series of holes in the case-hardened surface layer below which are voids. Such phenomena are common in the granitic terrain of western Corsica (Fig. 4.1b).

There is one other important interaction between mechanical and chemical weathering processes that results in specific landforms. This is an important process in granite terrain with well developed pressure-release jointing. Water penetrating along the cracks preferentially chemically weathers the rock in zones where the crack and joint density is highest. This leaves unweathered boulders (corestones), surrounded by "grus" a weak mixture of the weathering products of the granite (quartz, clay minerals, iron oxides). These are easily eroded to leave the corestones often in multiple layers as tors. There are tors on some of the granites of Brittany, and possibly in Limousin in the Massif Central. I have not seen tors there, but in several places I have seen corestones exposed in quarry faces. Chemical weathering appears to have been more intense during Pleistocene interglacials/interstadials, perhaps with frost shattering of the exposed tors during the intervening glacials/stadials.

The products of chemical weathering may accumulate *in situ* as a residual deposit, may be incorporated by biological processes into the soil (see below), or may if soluble, be removed by runoff or by percolating water.

Climatic Control of Weathering Processes—Soils Weathering processes, both mechanical and chemical are very much climatically controlled either directly, or indirectly through biology. When relatively fine-grained mineral debris is incorporated with biological material **soil formation** takes place. In most climatic situations (as in most of France) the dominant movement of moisture and any soil particles is vertically downwards within the soil, modified to some degree by biologically-based mixing. Over time this downward movement produces a vertically differentiated soil profile. However in climatic regimes where evaporation is a major process (dry climates: see above) there may be upwards movement of moisture and resultant precipitation of dissolved minerals at a horizon within the soil (forming a hardpan) or near the surface.

Different soil types develop under different environmental conditions, but whatever the case this is a slow process. Soil maturity is therefore a useful relative-age indicator on stable surfaces. For example on a suite of river terraces soil status can give an indication of the timing of the cessation of sedimentation, and therefore can be a useful aid to understanding their geomorphology. This is especially so when related to other methods of absolute age estimation (such as, for example radiocarbon dating).

Karst Geomorphology There is one special case of landforms and landscapes directly produced by chemical weathering processes that is particularly important in a French context, "karst" landscapes. The name karst is derived from an area in the Balkans/Slovenia. It is given to limestone landscapes dominated by solutional processes. Limestone (formed of calcite, $CaCO_3$) is subject to solution in rainwater (a dilute carbonic acid). The resulting surface forms on outcropping limestone are dominated by small-scale solutional features (rills, runnels, potholes) which together are referred to as limestone pavement (Fig. 4.1c). Surface solutional features are especially well developed in areas where previous glaciation had stripped the rock surfaces bare. Solution continues below the surface to create near-surface potholes (which may fill with other material—see Fig. 3.3d) or cave systems. Cave walls and ceilings may be ornamented with re-precipitated calcite features (tufa curtains, stalactites, stalagmites: Fig. 4.1d). Karstic features vary enormously in scale from small-scale surface solutional or depositional features to large scale surface depressions, gorges and canyons produced by a combination of fluvial activity, karstic solution and collapse (see later Figs. 8.6b, 8.6c, 14.2a, b). In addition rivers may disappear underground, only to re-emerge elsewhere as major springs, "Vauclusian springs" (named after the Vaucluse plateau in Provence).

Within France there are major karst areas on Jurassic limestones, especially in the Causses areas to the south of the Massif Central (see Chap. 8) and within the Jura mountains (see Chap. 12). On Cretaceous limestones there are major karst areas in the southern pre-Alps, in Provence, especially in Vaucluse, and in the Ardèche plateau to the west of the middle Rhône valley (see Chap. 14). For the distribution of karst areas in France see Fig. 4.2.

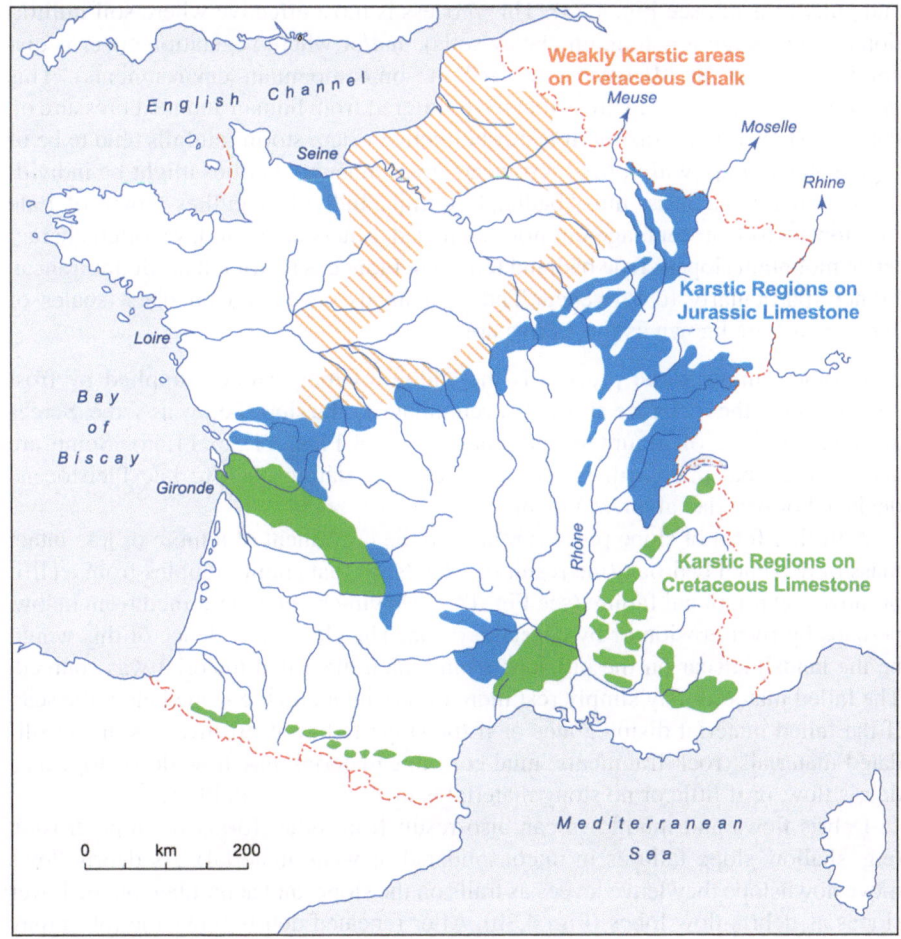

Fig. 4.2 Map of the main karst regions of France (Source: Vanara and Nicod 2013)

On resistant limestones where there has been tectonic uplift (e.g. around the southern flanks of the Alps, and the southern flanks of the Massif Central) many of the main rivers are deeply trenched in canyons. The best known are perhaps the Grand Canyon of the Verdon in Provence (see Chap. 14, see also Fig. 14.6c), the Ardèche Gorges (see Chap. 14, see also Figs. 14.2a, b) west of the Rhône, and the Gorges du Tarn in the Grand Causses area in the south of the Massif Central (see Chap. 8, see also Fig. 8.6b). In the north and east of France the Cretaceous Chalk plateaux show some karstic features such as solutional pipes (see Fig. 3.3d), disappearing drainage and strong springflow, but lack spectacular surface karstic features.

Slope Processes Slope processes involve the transfer of material downslope by a variety of mechanisms. The simplest process perhaps is where overland flow by runoff following heavy rainstorms, simply either transports loose material downslopes (sheet erosion) or incises into the slope to create rills and gullies (rill and gully erosion, see Fig. 4.3a). This process is most effective where soil infiltration capacities are low (e.g. on clayey soils), and/or where vegetation cover is discontinuous (e.g. in Mediterranean and in some mountain environments). This process is also effective in areas that have suffered from human-induced pressure on the land, such as overgrazing. It is most effective where storm rainfalls tend to be of high intensity (e.g. within Mediterranean environments). Gullies might be individual features or coalesce into "badland" terrain. Individual gullies (some of Late Pleistocene to Holocene age but now stabilised, others active today), often characterise mountain slopes. True badland terrain in France is restricted to Mediterranean France, particularly to Provence. There, badlands developed on black shales of Jurassic age are known as the "terres noires".

Another simple slope process is the sliding of loose stones supplied by frost weathering of the rock face above, to accumulate on the slope below as scree. Screes are characteristic of mountain environments (see Figs. 11.3c, 11.4a). Some are active today, but many others are relict features, dating back to late Pleistocene deglacial or periglacial environments.

A further form of slope process relates to the movement of a more or less intact mass of detached bedrock (the result of a landslip) that simply topples from a cliff, or moves by rotational failure (see Fig. 13.5c) having been undermined from below, perhaps by river erosion or by coastal erosion. The clearest evidence of this would be the landslide scar on the hillslope from which the failed material was sourced. The failed material may simply rest more or less intact on the slope below the scar. If the failed material disintegrates or if the slope failure itself involves unconsolidated materials (rock fragments, mud etc.) this material may flow downslope as a debris flow, or if little or no stony material is involved, as a mudflow.

Debris flows and mudflows can also result from other forms of slope erosion (e.g. shallow slope failures in unconsolidated or weak material). As debris flows move downslope they leave levees as trails on the slope, and acumulate on the lower slopes as debris-flow lobes (Fig. 4.3b). After repeated debris flows the lobes may coalesce as debris cones (small-scale) or (larger-scale) debris fans. (These are the end-members of fan-shaped depositional forms which include alluvial fans, fan

Fig. 4.3 Slope forms, including those related to Pleistocene periglacial processes. (**a**) Gully system: eroded by overland flow derived from runoff on the Upper Jurassic "Marnes Noires" in the Sainte Jalle badlands, Les Baronnies, northwest Provence (see also Fig. 14.3a). (**b**) Debris-flow lobes: on the flanks of Sambuy Mountain, south of Faverges, Bauges Massif, in the Pre-Alps of Savoie. Note the levées and lobe fronts. (**c**) Teracettes: small-scale indicators of active creep or other shallow mass-movement processes within the soil layer, on the flanks of the Grosne valley, near Cormatin, southern Burgundy. The image reflects the effects of the drought in the summer of 2018. (**d**) Patterned ground near the summit of Mont Ventoux, Provence: the result of seasonal surface freeze-thaw by Pleistocene permafrost. (**e**) Slope deposits at Calenzana, NW Corsica: The deposits at the top of the section are 'head' deposits, derived by mechanical weathering (frost shattering) from exposed bedrock further upslope, then transported downslope (by solifluction). Each of these processes would indicate cold, probably periglacial conditions. The lower part of the section comprises deeply chemically-weathered granitic material (indicative of warm humid climatic conditions—during an interglacial?) derived from underlying bedrock

deltas and deltas—see below: see also Figs. 4.6a, 11.3c, 11.4b). Within France, slope failures, landslides and debris flows are most characteristic of mountain environments. At the local scale their occurrence may reflect the local geological structure as well as topographic factors.

Much more ubiquitous on relatively steep slopes is creep, the almost imperceptible downslope movement of the soil or surface layers of unconsolidated material, primarily as the result of wetting and drying or freezing and thawing. Its most obvious expression is as terracettes (erroneously described as "sheep tracks") on steep grassy hillslopes (Fig. 4.3c). Such forms are characteristic of scarp slopes and steep valley sides throughout northern and eastern France.

Slope processes are much less active today than they were during the Pleistocene. Then, the occurrence of intense freezing and seasonal thawing and the widespread distribution of permafrost affected slope processes in all parts of France. In rocky or stony mountain environments "blockfields" were produced. Where there were mixed stony and fine soils differential freezing pressures within the mixed material laterally sorted the stones from the fines creating patterned ground, stone garlands on relatively flat slopes and stone stripes aligned downslope on steeper slopes. Pleistocene patterned ground is present at high elevations in the mountains of southern France (Fig. 4.3d). On even relatively gentle slopes the summer-thawed, wet "active layer" above the permanently frozen sub-soil often became unstable, sludging downslope to produce an ill-sorted slope deposit ("head") as a blanket, especially over the lower parts of the slope. Such deposits may bury older soils or slope deposits (Fig. 4.3e). Throughout northern and eastern France particularly, such superficial material is very common, especially mantling the lower slopes of escarpments.

Fluvial Processes Fluvial processes refer to erosion, transport and deposition by flowing water within channels. The term could include rill and gully channels (see above), but in this section I will be concerned with larger features, stream and river channels. Stream power is dependent on flood discharge and the channel gradient. Where stream power is high, especially where channels are steep, no sediment will accumulate, and channels will be erosional, scoured to bedrock (Fig. 4.4a, b). An extreme case is a waterfall (Fig. 4.4c), where there is an abrupt step in the channel floor, its location related probably to the longer term geomorphological history.

Where stream power is less than in *bedrock-scoured channels*, sediment may be deposited, resulting in a channel that can adjust its form by both erosion and deposition: an *"alluvial channel"*. Alluvial channel size will tend to reflect the magnitude of the normal range of flood discharges the channel carries; these are also times of maximum sediment transport, flood conditions that occur once or twice a year. Overbank discharges resulting in (usually minor) flooding would be expected on about the same frequency. Obviously more extensive and potentially disastrous flooding occurs more rarely. The cross-sectional shape of the channel within its floodplain will reflect both discharge magnitude and sediment calibre. Higher discharges result in relatively wider channels, as do coarser (sandier) sediments (easier to erode) as opposed to finer (more cohesive, clayey) sediments (more resistant to erosion).

In three dimensions, the channel pattern also reflects the environment. Some single-thread channels may simply be mildly sinuous and often of low gradient and relatively stable within clayey or vegetated banks. More common are meandering

Fig. 4.4 Fluvial features: on bedrock. (**a**) A simple bedrock channel within a relatively shallow gorge: Éyrieux Gorge, east of Le Cheylard, Auvergne. (**b**) A deeply incised bedrock channel: The Grand Canyon of the Verdon, cut into Upper Jurassic and Lower Cretaceous limestones, southwest of Castellane, Provence. (**c**) A valley-side waterfall on the margins of a (Late Pleistocene) glacial trough, fed by a hanging valley: Les Écrins, Northern Alps

channels where erosion is alternately concentrated on one bank with deposition on the other (Fig. 4.5a) resulting in the lateral migration of the channel position. The bed form in such a channel reflects the pattern, with deeper pools on the outside of the eroding bends and shallows (riffles) at the crossings between meander bends. Meanders migrate through time, renewing the floodplain as they do so. Single-thread meandering channels occur in relation to moderate floodflows on rivers whose sediment load includes at least some fine sediment.

On higher power rivers, carrying more coarse sand and gravel sediment, the non-cohesive nature of the deposited sediment allows excessively wide channels to develop, within which the simple flow patterns characteristic of meandering channels break down. The result is multi-cellular flows, which foster the development of within-channel sand bars and gravel bars. The flow separates into discrete sub-channels, leading to the development of braided patterns (Fig. 4.5b). Within France, braided channels occur on some of the larger rivers, especially the Loire (Fig. 9.12a–c), but are particularly characteristic of mountain and Mediterranean environments (Figs. 3.5, 4.5b, 13.5d, 14.6b, 15.4b).

There is another style of multiple channel, known as an anastomosing channel. This style used to be common prior to human-induced modifications, especially on low gradient rivers in northern France. Such channels differ from true braided channels, primarily because they are relatively narrow channels where the channel division is initiated by spillage (avulsion) from a discrete channel, rather than as a result of channel widening and flow separation within the channel. Anastomosing channels develop in low-gradient situations on rivers carrying fine (clayey) sediment. They are often associated with backswamps, levées and lagoons. They are

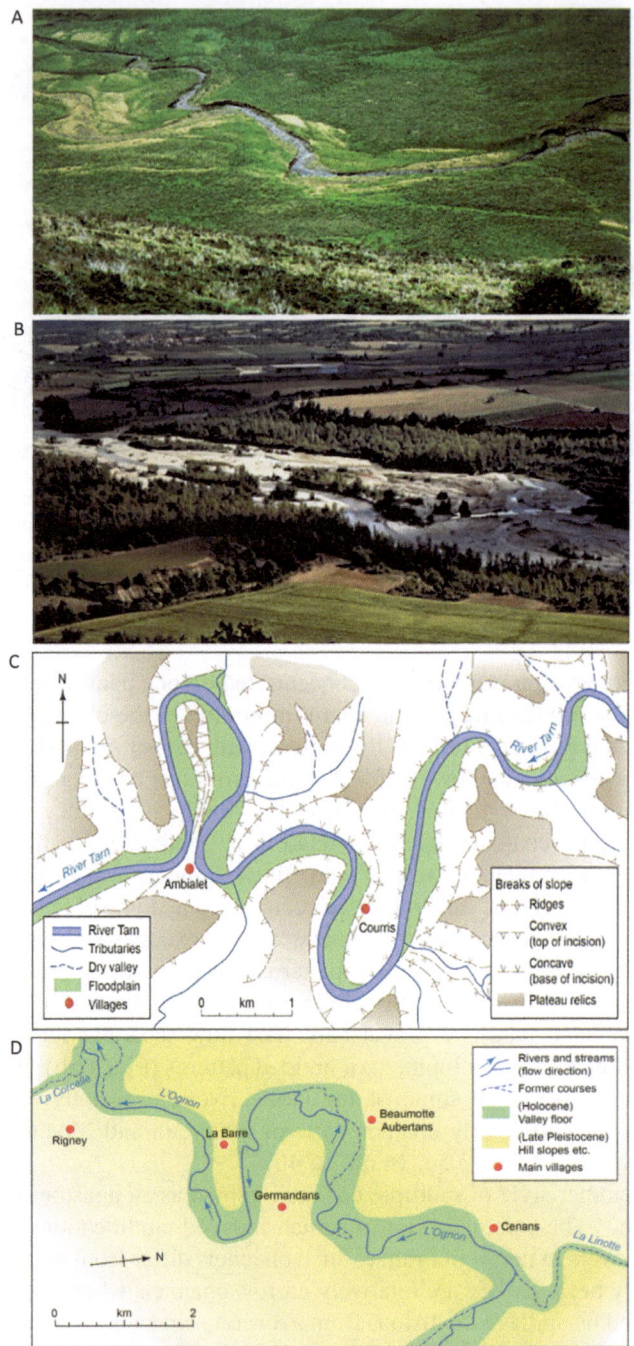

Fig. 4.5 Fluvial features: alluvial channels, including incised meanders. (**a**) A small meandering alluvial channel set within its own floodplain (one of thousands in France), showing evidence of recent (probably historic) channel change: the upper Santoine valley, Cantal, southern Auvergne.

characteristic of channels within deltas, but are also characteristic of the misfit modern channels within incised meandering valleys especially in northern France.

Incised meanders appear to relate to higher discharges during Pleistocene incision (see Chap. 3). In fact there are three characteristic forms of relationship between the incised meandering valley morphology and that of the channel itself within the meandering valley. In some cases the meander geometry of the modern channel mirrors that of the incised meanders (Fig. 4.5c, 14.2b) implying either bedrock control or perhaps, little drainage contraction since the Late (?) Pleistocene. Secondly, the modern meandering channel may be markedly "misfit" within the incised meandering valley (Fig. 4.5d) implying a marked reduction in flood discharge since the valley incised. Thirdly, the modern channel may also be markedly clearly underfit but characterised by an anastomosing rather than a meandering channel. That would imply an even greater discharge reduction than would simple underfit channel meanders.

Fluvial deposition produces a variety of forms. On a main valley floor alluvial sediment deposited by the river itself (both by braided and by meandering rivers) forms the floodplain. Over time the floodplain tends to be destroyed and rebuilt in relation to channel migration. Anastomosing channels however, may tend to be associated with net deposition (albeit slowly), ultimately building up the elevation of the valley floor.

In mountain areas where steep sediment-laden tributaries enter a lower gradient main valley, they may deposit their sediment load in the form of an alluvial fan (Fig. 4.6a). The fan may trap most of the sediment derived from the tributary catchment, in which case it "buffers" the system. Alternatively if it is trenched through, or eroded by the main stream, it may become "coupled", linking the tributary sediment source with the main catchment. The location of alluvial fans may simply be topographically controlled (tributary-junction settings as described above e.g. Figs. 4.6a, 11.4b), or they may be structurally-controlled along a fault scarp for example or along a faulted mountain front. They may be related to modern pathways of sediment movement and be wholly Holocene in age. Alternatively, they may be related to pathways of sediment movement during the last glaciation and therefore be of Pleistocene age. Alluvial fans may be much larger features, "megafans", related to longer-term structural and geomorphic evolution, for example the Neogene-age Lannemezan megafan fronting the Pyrenees (Fig. 4.6b, see also Chap. 11).

Other forms of fluvial deposition relate to the fluvial-lacustrine or fluvial-marine interfaces. Mountain rivers entering small lakes may deposit their sediment load at the lake margin as a fan-delta (see Fig. 11.3c). On the other hand large sediment-laden rivers may build up a delta at the coast; classic in France is the Camargue (the

---

Fig. 4.5 (continued) (**b**) A gravel-bed braided reach of the Buëch River, northern Provençal Alps (see also Fig. 14.4 map). (**c**) Map of simple incised meanders of the River Tarn, near Ambialet, southwestern margins of the Massif Central. Note the more or less similar geometry of the valley meanders and the channel meanders. (**d**) Map of incised meanders of the River Ognon, south of Rioz, northern margin of the Jura mountains, Franche Compté. Note the markedly misfit relationship of the modern channel meanders within the much larger valley meanders

Rhône delta, see Chap. 14). More commonly, the seaward end of what had been extensions of Pleistocene river valleys became drowned by the post-glacial rise in sea level to form estuaries (e.g. the Gironde estuary, Aquitaine: Figs. 10.2c, 10.3 map). Most rivers on the Channel and Atlantic coasts of France terminate in estuaries. These rivers carry lower sediment loads than the Mediterranean rivers. Also the effects of tidal scour on the (tidal) Channel and Atlantic coasts tend to prevent excessive deposition, thus maintaining open water in the main channel of an estuary. Deltas may also occur at the fluvial-lacustrine interface.

Within a river valley the quasi-equilibrium of an alluvial channel may be interrupted by an environmental change that results in one of two possibilities. On the

Fig. 4.6 Alluvial fans. (**a**) A small alluvial fan (what is effectively a tributary-junction alluvial fan): Col des Aravis, northern Alps. Note: The fan channel incised within a fanhead trench in the upper part of the fan, but that channel is on the fan surface in the distal part of the fan, and depositing sediment there. (**b**) Landsat image of the Lannemezan megafan [E 0.24′ N 43.06′] on the northern margin of the Pyrenees: Initially of Miocene age, it has been dissected since then. Note the radial nature of the distributary channels across the fan surface. Note also the transverse drainage (rivers crossing folded structures) to the SE (bottom right of the image)

one hand an effective increase in sediment load and/or a decrease in flood power would result in excess sedimentation and the eventual burial of the former flood-plain and valley floor. On the other hand the reverse would lead to excessive erosion, probably also involving incision. The climatic changes of the Late Pleistocene and the climatic and human-induced changes of the Holocene resulted in switches between one condition and the other. The river terraces (see Fig. 3.7c) that charac-terise the valleys of many French rivers resulted from these sequences. High sedi-ment loads of the late Pleistocene caused aggradation, followed by reduced sediment loads during the Holocene, causing incision. The other major post-Pleistocene adjustment (already mentioned above), relating to the incised meanders on many French river valleys, was the channel shrinkage to produce misfit modern meanders within much larger meandering valleys (see maps Figs. 3.5, 4.5d).

A final cautionary word: the river systems of France, particularly the larger rivers, have been substantially modified by human activity (see Fig. 3.8 map). This is espe-cially true of the Rhône, the Seine, and the Rhine (in so far as that river is French), less so of the Loire and perhaps the Saône, although the Saône has weirs and locks for navigation. The Rhône is essentially an artificial river, modified by dams and diversions built primarily for hydro-electric generation. The Seine has been substan-tially modified by bank protection and weirs and locks provided for navigation. The Rhine, at least where it forms the French border, is essentially an artificial channel. Many of the other lesser rivers are at least partially natural and provide an opportu-nity for appreciating the diversity of French fluvial geomorphology.

Coastal Processes France has a diverse range of coastal landscapes along the Channel, Atlantic and Mediterranean coasts and around the island of Corsica. These landscapes are young. Post-glacial global sea levels attained their present elevations only about 7000 years ago, though there are in places landforms and sediments cre-ated by the previous high sea levels of the last interglacial around 100,000 BP. Locally the modern coasts are dominated by either erosion or by deposition. The factors that affect the modern coastal morphology are (i) the landward terrain, including its geology, (ii) the offshore topography, especially the shoreface gradi-ent, and (iii) the marine dynamics, including tidal range and exposure.

Erosional coasts (Figs. 1.3c, 4.7a, 6.3a, 14.8a, 15.5b) occur where landward ele-vations are sufficient for cliff development. On such coasts cliffs dominate, although pocket beaches may be present especially at the mouths of small valleys. The cliffs themselves may be footed by beaches or by bare erosional wave-cut platforms. In outline, coastal irregularity reflects the complexity of the bedrock geology. Contrast the straight Chalk cliffs of Normandy (see Chap. 9, see also Fig. 1.3c) with the complex plan view of the granitic cliffs of Brittany (see Chap. 6, see also Fig. 6.3a). Other features include erosional stacks (Figs. 1.3c, 4.7a, 6.3a) remaining after cliff erosion and recession, and in some bizarre cases, natural arches (Fig. 1.3c) again products of differential erosion. Remnants of former erosional coasts may be pre-served as rock platforms (Fig. 3.7d), somewhat above modern sea level or at higher elevations.

Depositional coasts tend to occur where both seaward and landward gradients are low. Such coasts include a great variety of form (Figs. 4.7b, 6.3b, c). Beaches themselves are dominated by sand (Fig. 4.7b) and gravel sized sediments, but finer sediments (silts and clays, Fig. 6.3c) may occur in other depositional settings e.g. estuaries (Fig. 6.3c), deltas, back lagoons etc. (Fig. 14.11). Beaches may be linear features or convolute forms involving complex spits growing into deeper water. Spits may have landward-pointing recurved laterals that relate to previous spit positions. With spit growth the landward zone becomes increasingly sheltered from wave action and may become a site for salt-marsh development.

Complex depositional forms are often associated with estuaries and deltas. Estuaries are drowned fluvial valleys at the outlet of river systems drowned by the post-glacial rise in sea level. To some extent they may be kept relatively sediment free by tidal scour. Within estuaries however, there are likely to be low-energy locations where fine sedimentation may occur and subsequently salt marshes may develop.

Deltas also have developed in their present form since the post-glacial rise in sea level and are more likely where sediment-laden rivers enter the sea, especially where there is little or no tidal scour—both conditions more likely on the

Fig. 4.7 Coastal landforms. (**a**) Complex cliffs cut in fractured Hercynian granite: Nez de Joburg, Cotentin peninsula, Normandy. Note the offshore stacks. (**b**) Extensive intertidal sand beach: Carteret, west coast of the Cotentin peninsula, Normandy. Note the partially buried rock platform at the head of the beach. (**c**) Pilat dune, Landes Coast, Aquitaine: Shutterstock image 209,857,321 *(copyright: Alberto Loyo, Shutterstock 209,857,321)*

Mediterranean than on the Atlantic or Channel coasts of France. The "delta of all deltas" is the Rhône delta—the Camargue (see Chap. 14).

The present coastal features are obviously dynamic and have been developing at their present elevations for only seven thousand years or so. Over that period relative sea level may not have been static, and has been subject to minor eustatic variations, as well as to local tectonic and isostatic controls of the land elevation (e.g. Regnauld et al. 1996; Vella and Provansal 2000). Sea level is not the only variable factor. Sediment availability and not least human activity (e.g. coastal protection measures, reclamation etc.) have varied over the same period. The coasts are dynamic and are subject to rapid natural and human-induced change.

Aeolian Processes Aeolian processes are not particularly important in the overall geomorphology of France except in two cases, both associated with other process domains. In the first case, aeolian activity is an important coastal process (Fig. 4.7c). In areas of high tidal range, low shoreface gradient and dominant onshore winds, sand can be blown onshore accumulating in dunes. The enormous dunefield along the Landes coast in southwest France (see Fig. 10.3 map) from the Gironde estuary almost to the Spanish border includes the dune "Pilat", the largest dune in Europe (110 m high) (Fig. 4.7c).

In the second case, during Pleistocene periglacial conditions (see above) wind-blown silt (loëss), derived from glacial margins further north, was deposited across the North European plain, including northern France. It made little difference to the topography, but was important from a human point of view in adding fertility to the soil and in improving soil structure.

Glacial Processes Glacial processes depend on glacial form, which can range from continental-scale ice sheets to small patches of ice on mountain slopes. During the Pleistocene France was not directly affected by continental-scale ice sheets as were Britain and northern Europe, but had regional mountain ice caps over the Alps/Jura, a smaller cap over the Pyrenees, mostly small cirque glaciers in the Vosges and Massif Central (see Fig. 3.6 map), and Corsica (Conchon 1978, 1986; Calvet and Gunnell 2008; Kuhlemann et al. 2005). Today cirque and valley glaciers persist in the Alps (see Fig. 3.6 map) notably in the Mont Blanc massif (Fig. 4.8a; see also Fig. 13.3 map, Figs. 1.3a, 13.4b, c). Modern glaciers also occur in the Vanoise, Écrins (see Fig. 13.4a) and Queyras massifs, in those areas mostly as cirque glaciers. In the Pyrenees there are small cirque glaciers and ice patches (see Fig. 11.3a). All the modern glaciers are currently shrinking.

Glaciers, especially mountain cirque and valley glaciers, are very powerful agents of erosion. They are driven by gravity and unlike rivers they are not limited by base-level controls. Armoured by rock debris at their base they are capable of scouring deep valleys, which on deglaciation form troughs. Within mountain areas, glaciers are often sourced in deep near-circular basins, which on deglaciation are preserved as cirques (see Figs. 1.3b, 3.7a). Active cirques are present in the Alps and the Pyrenees. Pleistocene cirque basins are also present in the Vosges and the Massif Central. Similarly on deglaciation, tributary valleys may be "hanging valleys"

(Fig. 4.4c) perched above the main valley floor. Also on deglaciation, lake basins may be abundant in previously glaciated terrain (see Figs. 12.4c, 13.5a).

Glaciers carry enormous amounts of ill-sorted sediment, which they dump on melting, forming moraines (Fig. 4.8b). This may simply be "ground moraine" at the base of the ice and dumped in an irregular sheet, or may be specific morainic ridges dumped at the ice margin, often responding to seasonal acivity. On deglaciation, such ridges may enclose lake basins, as is the case of a number of lakes in the Jura associated with the Last Glacial limits (see Chap. 12; see also Fig. 12.4c). Successive morainic ridges may also mark the recessional stages of glacial melt.

Glaciers involve not only ice but glacial meltwater, which is also a powerful agent of erosion, both extra-glacially and sub-glacially. Many so-called "glacial drainage diversions" where the preglacial drainage pattern was radically altered by glaciation, may owe as much to sub-glacial meltwater erosion (creating "meltwater channels") as to erosion by ice itself. Extra-glacially meltwater rivers carry very high sediment loads, often dumped as a valley-filling gravel spread (known as a "sandur") immediately in front of the ice limit.

In addition to the direct effects of Pleistocene glaciation itself, during glacial maxima most of France was affected by periglacial processes, described above under "slope processes".

A   B

Fig. 4.8 Glacial geomorphology. (**a**) Landsat image of a large valley glacier, Mer de Glace [E 7.00′ N 45.56′], Mont Blanc range. Note the arcuate annual(?) banding of clean and dirty ice. Note also the marginal and terminal morainic zones, indicative of the recent shrinkage of the glacier. (**b**) Pleistocene moraines, St Laurent, Jura

Influence of Climate on Modern Geomorphic Processes Today France has three major climatic zones (Atlantic, Continental, Mediterranean), each with a distinct influence on geomorphic processes, with broad transitional zones bounding the core areas. The climate within each zone varies to a greater or lesser degree with elevation.

The Atlantic climatic zone affects western and northern coastal areas, and inland into Aquitaine, and the western part of the Paris basin. The weather is dominated by maritime temperate air masses with relatively mild seasonal temperatures and abundant moisture. Depressions and frontal systems are important. Overall, rainfall occurs throughout the year but snow is rare. Winter temperatures are mild. Frosts do occur, but less frequently and less severely than further inland. Summers tend to be cooler than in the rest of France, but there is an overall temperature gradient increasing southwards. There is a transition inland towards continental trends and in the far south, eastwards towards Mediterranean trends. In the far south the Pyrenees have their own mountain climate, again with an increase in continentality eastwards. Geomorphic processes in this zone are water-balance dominated, therefore there is a winter flood dominance.

The Continental climatic zone is most pronounced in eastern France. It is dominated by interactions between continental and maritime air masses. There is abundant moisture and as in the Atlantic zone, depressions and frontal systems are important, but the continental influence is much greater. In winter it is colder than in the Atlantic zone, with frequent frosts, and snow especially at higher elevations. It can be hot in summer with thunderstorms. There are obviously transitions: westwards towards a more maritime influence; southeastwards there is an increase in continentality, and with increasing elevation into the Jura and the Alps, there is an increase in winter snowfall; southwestwards there is also the elevational effect into the Massif Central. Southwards down the Rhône valley there is a transition towards the Mediterranean zone. Within the continental climate zone geomorphic effectiveness is expressed by late winter/early spring maximum runoff, plus the snowmelt effect especially from high elevations. Summer thunderstorms may be important locally.

The true Mediterranean climatic zone, usually with a marked seasonality of hot dry summers and mild damp winters, is restricted to the Mediterranean coastal areas. There is a transition to greater continentality northwards into northern Provence and the southern Alps, and northwest into the margins of the Massif Central. Westwards there is a transition towards the Atlantic zone through the Carcasonne Gate into Aquitaine. Geomorphic effectiveness is expressed by moderate runoff in winter, especially in the transition zone into northern Provence, with an early spring runoff maximum. Occasional summer thunderstorms, especially in late summer, may be important locally.

Chapter 5
French Cultural Landscapes: Relation to Geomorphology

The Gross Historical Context

At the grossest scale there is a broad relationship between the terrain and the historical development of France as a nation. Think of the nature of the French borders. The southern and southeastern borders are more or less defined by the Pyrenees, the Alps and to a lesser extent the Jura. Within the last two hundred years there has been relatively little military conflict across these borders. This is in complete contrast with the repeated conflicts across the northeastern and eastern borders, across terrain over which military movement was much simpler.

Prior to the Middle Ages the territory of France that we recognise today hardly existed. France became established as a small kingdom in the twelfth Century. The Middle Ages were dominated by conflict with the English throne, which from 1154 to 1286 held most of west and southwest France as well as Normandy. This was broadly the position until 1453, the end of the 100-Years War, when France regained virtually all of the English possessions in the west and southwest. Burgundy was independent, and incidentally extended way beyond the present frontiers of France, north into the Netherlands. In 1477 Burgundy (excluding the Netherlands) was annexed by France but there continued to be intermittent conflict between France and Burgundy. In 1678 under Louis XIV Franche Comté was annexed and France became an integral nation, with most of its north-eastern border more or less established. The eastern border however, particularly involving Lorraine and Alsace, continued to be disputed until the end of the second World War.

During the nineteenth and twentieth Centuries the nature of the terrain in the east continued to influence international conflict, allowing the free movement of armies during the Napoleonic wars, the Franco-Prussian war and the World Wars of the twentieth Century. Following the Franco-Prussian war (from 1871) Alsace and Lorraine became German but were returned to France after the First World War

© The Author(s), under exclusive license to Springer Nature
Switzerland AG 2025
A. Harvey, *The Geomorphology of French Landscapes*,
https://doi.org/10.1007/978-3-031-68490-6_5

(1918), only to be ceded to Germany again in 1940 to be returned finally to France in 1945.

Rural Landscapes

At the much more focussed "landscape" scale than that described above, and having considered the main processes creating the physical landscape (see Chap. 4), we are in a position to return to a theme introduced in Chap. 1, that of the relationships between the physical and human landscapes within France. There are numerous superb books on French rural landscapes, often with beautiful photos, but lacking in any real geological or geomorphic basis. Their emphasis is often on villages within the landscape.

The first obvious characteristics of French villages are the traditional building materials used and the building style. The first aspect, building materials, does reflect the regional geology (but only indirectly the geomorphology); that is the local availability of building stones. In areas where suitable building stone was not locally available a characteristic rural building style is the half-timbered frame with brick infill. This style characterises the Chalk areas of northern France, (see Chap. 9), the Quaternary Bresse plains (see Chap. 12), and the Tertiary centre of the Aquitaine basin (see Chap. 10). The second aspect, building style, is largely a regional cultural characteristic, again not directly related to the geomorphology. In building styles there is a broad contrast between the steep roofs characteristic of northern France and the low angle roofs characteristic of the south. This might be in part a cultural phenomenon, and in part be environmentally influenced, related to rain in the north (steep roofs), and sun in the south (low angle roofs providing shade where the roofs overhang beyond the walls). In the east the transition occurs in southern Burgundy; in the west it occurs in Poitou, between the Loire valley and northern Aquitaine. These transitions are also associated with different roofing styles and materials: "Roman-style" curved tiles in the south, flat tiles in the north. There are other regional roofing materials, for example the "lauze" basalt tiles of the volcanic districts in Auvergne, and slates in Brittany.

The physical landscape (particularly in relation to the patterns of land use) is of fundamental importance in how we perceive its cultural associations. This is true at both regional and local scales. Below I give three brief examples.

As a first example I cite the formerly characteristic "bocage" landscape of small grass fields between earth banks. Scattered stone-built farmsteads with steep roofs occur within sometimes deep valleys below plateau surfaces. At the regional scale the formerly bocage type of landscape coincides with the Hercynian "Armorican" massif dominated by metamorphic rocks, truncated by Neogene erosion surfaces (see Chap. 6). At a more detailed local scale, especially in relation to settlements within valleys incised into the plateau surfaces, we would need to consider the local effects of Late Pleistocene (periglacial) processes on the landforms and how the settlements relate to the local topography.

As a second example, consider the villages within the plains and scarplands of eastern Champagne or Lorraine (see Chap. 9). Within a landscape of huge open cereal fields such villages, inherited from the medieval village and open-field system, are often street villages. They often have a morphology of houses set at right angles to the main street, and built of local stone or are half-timbered with brick infill. At the regional scale the particular site is important (eg. plateau-surface, scarp foot etc). At a more detailed scale the relation to late Pleistocene periglacial landforms is important.

As a third and contrasting example, consider the hilltop or valley-side mountain villages within the Provençal Alps often situated on defensive sites (see Chap. 14). At the regional scale their relation to geological structure and rock outcrop is important, but at the local scale their situation in relation to Quaternary landform features is important too. Often they have traditional local small-scale closely cropped fields, perhaps with vines, but in many situations the relation to springs and water courses for small-scale irrigation may be important.

In terms of land use (and crop cover), we must realise two things: (i) almost all the land surface of France is "managed"; (ii) the situation is dynamic, changing with the economic and social climate.

Forest is a major land-cover type, mostly deciduous woodland in the north and in most lowland areas, with conifers mostly at higher elevations in the major mountain areas. Most forest areas are managed either for timber production or for fuel wood. Since the nineteenth Century there has been an increase in forest area, some, especially in the mountain areas, partly as a result of regrowth, but also as a result of replanting. Compare old photographs within the Alps and Pyrenees with the picture today. Within the Southern Alps (see Chap. 14) there has been a recent geomorphic response to reforestation, with a reduction of sediment supply to the streams resulting in channel narrowing, a reduction in braiding, and in extreme cases, scour of the stream bed to bedrock. There has also been systematic planting, especially of conifers, not only in the mountain areas, but in some lowland areas, for example in the Landes plains in Aquitaine (see Chap. 10).

The major land cover, especially in the scarplands and plateau areas in northern and eastern France, is arable land. The major crop is wheat—think of how important bread is in the traditional French diet! Over the last few decades there has been some diversification, for example into root crops, corn, sunflowers and rape seed. Balancing the arable land there is extensive pasture, especially in areas of wetter climates in the west, and "wetter" soils in the major valleys and in the uplands, for example in the Massif Central. Grazing land is primarily for cattle—e.g. for beef cattle, Charollais and Limousin breeds especially, and for dairy cattle. Cheese is a hugely important part of the French diet. Think of De Gaulle's 1962 statement "How can anyone be expected to govern a country with 325 cheeses?"—almost certainly an under-estimate! Goats are important and contribute significantly to cheese production, but tend to be reared on the farm itself and grazed near the farm. Sheep are much less important than in Britain, though there are some in marginal mountain areas in the south, and in some estuarine areas in the north.

Wine

Wine is one of the most important French products at both the national level, and at the local level in the wine producing regions. Vine cultivation also produces very distinctive landscapes. Furthermore, within the context of overall climatic control, geology and especially geomorphology are very significant local factors for vine cultivation (see Pomerol 1989).

In most of the more northerly wine-growing areas the vineyards are primarily on scarp slopes or valley sides (Fig. 5.1). They are found particularly on the mid- and lower slopes, especially on Pleistocene colluvial material below the escarpments themselves. In Champagne the scarp is that of the Eocene limestone (the Calcaire de Brie) that forms the Falaise de l'Île-de-France to the southwest of Rheims, and forms the slopes above the Marne valley near Épernay (see Fig. 9.6a). In Alsace the vineyards are concentrated on colluvial material in the fault zone on the margins of the Vosges, southwest of Strasbourg (see Figs. 7.3b, 7.4 map). In Burgundy the vineyards are on colluvial material below the Jurassic limestone scarps of the Côte d'Or, Côte de Nuits, Côte de Beaune, Côte Chalonnaise (Fig. 5.1), and the Mâconnais ridges (see Fig. 12.6a). The Beaujolais vineyards (see Fig. 12.6c), south of Mâcon are on the colluvial lower slopes of the granitic rocks of Beaujolais. In the Alps and the Jura the vineyards tend to be located on the lower slopes below limestone ridges.

Fig. 5.1 A vinyard village: Saules, Burgundy. The village is situated on the Côte Chalonnaise escarpment, capped by Middle Jurassic Limestones. This view is across the Grosne valley, beyond which are the Mâconnais hills, also dominated by folded Middle Jurassic limestone—more wine terrain in that area

For example in southern Burgundy, the wine village of Saules (Fig. 5.1) on the Côte Chalonaise lies at the foot of the fault-line escarpment of the Middle Jurassic limestone. To the south of the downfaulted Tertiary clays of the Grosne valley are more folded and faulted Jurassic rocks which form the Mâconais hills. During the Pleistocene, frost shattering of the limestone of the Côte Chalonais was followed by solifluction which transported much calcareous material from the face of the escarpment onto the slopes below. This resulted in soils highly suited to vine cultivation.

South of Lyon, the Côtes du Rhône vineyards are on colluvial slopes on the flanks of the incised Rhône valley. Further south still the Provençal vines are much more widespread, as are those of Languedoc/Rousillon. This greater spread is probably in response to climate. With fewer winter frosts here the need for hillslope frost drainage is probably less. In Aquitaine there are valley-side vineyards in the inland areas of Bergerac, Cahors and Agen. Nearer the Atlantic coast in the Bordeaux, Médoc, Graves, Sauternes and Entre-deux-Mers areas (Fig. 10.3 map), also further north in the Charente and Cognac (brandy) areas, the vines are widespread over a range of terrains. This pattern is repeated in the Loire area, with inland vineyards in the Sancerre and Orléanais areas restricted largely to the valley sides, but towards the milder oceanic climates of Anjou, Saumur and Nantais the vineyards are more extensive, over a range of different terrain types.

It is now time to move on to consider the geomorphology of the regions of France.

Part II
The Geomorphic Regions of France

The second part of this book deals with the geomorphology of the regions of France, regions defined by their geomorphology (See Fig. 1.1) not by their human history nor as modern administrative regions. It is organised into three groups of chapters on the basis of the dominant geomorphic themes. The three groups are (i) The Hercynian massifs, relatively elevated regions dominated by plateaux mostly on Palaeozoic rocks; (ii) Intervening largely basinal areas of relatively subdued relief, mostly on Mesozoic and Cenozoic rocks; (iii) "Alpine" regions of mountains and adjacent areas dominated by complex Cenozoic structural deformation. Individual chapters present the elements of the regional geomorphology followed by my perceptions of the highlights.

These three chapters deal with the various Hercynian massifs: Armorica/Brittany (Chap. 6); The eastern massifs from the Ardennes to the Vosges, mostly along the French border with Belgium, Luxemburg and Germany (Chap. 7), and finally the Massif Central (Chap. 8). These areas have a number of geological features in common. The major structures in all three areas date from the Hercynian orogeny (culminating in the Permian) that resulted from the collision of the southern Gondwana continent with the northern Laurentia/Baltica continent. All three massifs have core zones of deformed Palaeozoic rocks showing varying degrees of metamorphism, from low-grade metamorphism in the Ardennes to high grade metamorphism in many parts of Armorica. All areas except the Ardennes, of which only a small part is in France, have major zones of granitic intrusions. The absence of granitic intrusions in the French Ardennes also applies to the eastward continuation of the Ardennes into Belgium (and Germany). Armorica particularly, but also the Vosges, is structurally complex, the Ardennes less so. The Massif Central is geologically complex in that it also includes younger rocks (Neogene-Quaternary volcanic rocks), plus it has a greater structural complexity as a result of modification by 'Tertiary'/Alpine tectonics. All these areas have something in common with each other in terms of landforms. All are relatively elevated in relation to the surrounding areas on Mesozoic or Cenozoic rocks (maximum elevations are approximately: Armorica 380 m; Ardennes 450 m; Vosges 1424 m; Massif Central 1855 m). All three areas have extensive areas of Neogene (?) erosion surfaces that truncate the underlying structures. The modern drainage is incised into these surfaces to a greater or lesser degree. The topography of the Massif Central is more varied and more complex than that of the other areas.

Chapter 6
Armorica (Brittany, Vendée, Cotentin)

Armorica includes the whole of Brittany, plus western Normandy including the Cotentin peninsula. It also includes the Vendée, south of the Loire estuary (Fig. 6.1 map). Brittany is often likened by English people to Devon and Cornwall. There are some similarities related to the geology and the geomorphology, but Brittany, which is only part of Armorica, is enormous and geologically much more complex than Devon and Cornwall.

The Geology of Armorica

Armorica forms the core area of the Hercynian mountain system whose formation culminated during the Permian. The dominant rocks are metamorphics, dating in part from the Lower Palaeozoic including some rocks affected by Caledonian metamorphism and structural deformation. There are also Upper Palaeozoic rocks, some of which were directly affected by Hercynian metamorphism and structural deformation. Finally, there are numerous Hercynian granitic intrusions across Brittany and south of the Loire in Vendée. In addition to the main Hercynian fold and local fault structures, there are major Late Hercynian NW-SE orientated strike-slip faults. These run sub-parallel to the southern coast of Brittany from the Pointe du Raz in the northwest right the way southeast into Vendée. Note that these fault alignments are picked out by the drainage (Figs. 6.1 map, 6.2b map).

Inland, the flanks of the Armorican Massif are overlain unconformably, primarily by the Jurassic rocks of the western margins of the Paris Basin, but also locally near Angers by Upper Cretaceous rocks.

A. Harvey, *The Geomorphology of French Landscapes*,
https://doi.org/10.1007/978-3-031-68490-6_6

Fig. 6.1 Map of Armorica

The Geomorphology of Armorica: Inland

The scenery of inland Brittany is rather dull and monotonous (Fig. 6.2a) comprising low-relief plateaux, degraded erosion surfaces that truncate the underlying Hercynian structures. These surfaces are probably Neogene in age, although in some areas they have been interpreted simply as stripped sub-Jurassic unconformity surfaces. I consider that unlikely partly because the angles seem to be wrong (near horizontal erosion surfaces, dipping sub-Jurassic surface). The surface form appears to relate to the likely pre-Pleistocene patterns of drainage evolution. Adding to the monotony of the upland landscape (to English eyes perhaps?) is the virtual absence of semi-wild "open moorlands". There are two small areas that are exceptions. One is the Monts d'Arrée (south of Morlaix, in northern Britttany), a small open moorland area, which is crowned by granite tors and includes the highest point in Brittany at only 380 m. The other, the Montagnes Noires (Fig. 6.2a), south of Carhaix-Plouguat, is an even smaller area and barely reaches 290 m elevation. Elsewhere the upland landscape is a farmed 'bocage' landscape of small fields with earthen hedge-banks devoted particularly to grazing, and rarely reaching an elevation of 200 m.

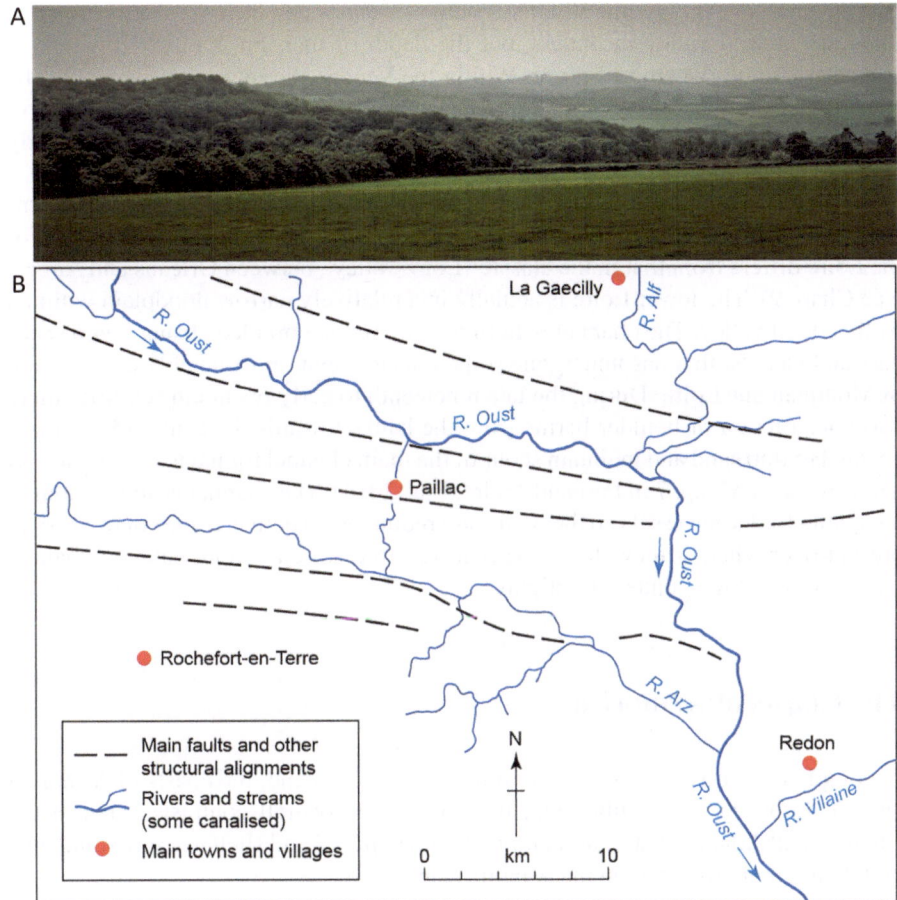

Fig. 6.2 Armorica landscapes: inland. (**a**) Level skylines of the Montagnes Noires, near Gourin: Neogene (?) erosion surfaces bevel the complex structures in the underlying Hercynian metamorphic rocks and granites of the Armorican massif. (**b**) Map of fault-aligned drainage near Redon, southern Brittany

There is little local relief even in the areas where the Hercynian rocks are overlain by Mesozoic rocks. At best there are only low escarpments in Jurassic limestones (e.g. east of the Sarthe valley, near Sablé-sur-Sarthe). Some variety is provided by the drainage, but mostly this is by short small streams in shallow valleys below the plateau surfaces. Locally, some pattern relates to drainage alignment with major structures, particularly the NW-SE strike-slip fault system (e.g. River Goyan, west of Quimper; Rivers Loc'h, Arz, and Liziec—all north and northeast of Vannes; the River Oust, northwest of Redon (Fig. 6.2b); and in the Vendée area south of the Loire, part of the River Sèvre Nantaise).

The river and stream channels themselves tend to be low gradient, single thread, meandering channels. On the larger rivers (e.g. The Aulne, east of Châteaulin; the

Blavet, south of Pontivy; the Villaine, south of Rennes and the rivers in the Vendée) there are incised valley meanders, but the depth of incision is rather less than is characteristic elsewhere in France. Within the Vendée the incised meanders of the Sèvre Nantaise have a valley and channel morphology suggesting misfit modern meanders within larger valley meanders—but this morphology is only poorly developed.

The only "large" river in the region is the Armorican section of the Loire (downstream of Angers to the tidal limit and the estuary beyond Nantes). This section of the Loire differs from that in the classic "Loire valley" between Orléans and Angers (see Chap. 9). The lower Loire is actually in a relatively narrow floodplain within a well defined valley. The channel pattern is of a sandbed braided channel with sandbars and islands, that has undergone some management, most notably downstream of Montjean-sur-Loire. During the late nineteenth to early twentieth centuries there was construction of boulder berms from the banks towards the centre of the main channel to trap sand and maintain scour of the main channel for navigation (Carcaud and Davodeau, Chap. 7 in Fort and André, Eds, 2014). The channel is now no longer navigable for large vessels so the berms are redundant. There is some controversy at present over whether they should be removed to restore a "more natural" channel (see section below—under Highlights).

The Coasts of Armorica

In complete contrast with the rather dull landscapes inland, the coasts of Armorica and Cotentin are spectacular (Fig. 6.3). There are overall contrasts between the Channel and Atlantic coasts in relation to both exposed and sheltered environments, both being more extreme on the Atlantic side.

The Channel Coast In the east is the Cotentin peninsula. This has a more gentle eastern (leeward) face than western (windward) face. The south end of the eastern face is primarily a depositional coast (the Carentan estuary and saltmarshes, north of which is one of the second World War Normandy landing sites: Utah Beach). The north coast of the peninsula is mostly an erosional rocky coast, culminating in its NW tip at Cap de la Hague/Nez de Jobourg, spectacular headlands cut in granitic and metamorphic rocks (Fig. 4.7a). There are tors on the clifftop at Nez de Jobourg. The west coast of the peninsula is a mixture of large sandy beaches (see Fig. 4.7b) and rocky headlands. Some of the beaches, such as that north of Carteret, are backed by dunes. South of Carteret the coast is primarily depositional, featuring wide sandy beaches, sand spits, lagoons and estuarine saltmarshes (Fig. 6.3b). The peninsula as a whole preserves a wide range of evidence of earlier sea levels, including higher level erosion surfaces (dating from pre-Mid Pleistocene times), and a set of lower and younger erosional platforms. Research on these forms (Regnauld et al. 1996) indicates an interaction between slow regional uplift and more rapid glacial/inter-

Fig. 6.3 Armorica, coastal landscapes (see also Fig. 4.7a, b). (**a**) The Pointe de Van, complex granites have been eroded to form an irregular cliffed coast. Note the offshore stacks. (**b**) The inlet at Port Bail on the west coast of the Cotentin peninsula. Away from the cliffed, exposed coasts much of the Armorica coast is depositional. (**c**) Mont St Michel, the isolated granodiorite boss within the Baie du Mont Saint Michel, eastern Brittany. This bay is the open coastal zone with the highest tidal range in France, a low energy depositional coast characterised by extensive salt marshes, a very small segment of which is shown in this picture! Shutterstock image 679,580,596 *(copyright: 4Max, Shutterstock 679,580,596)*

glacial sea-level oscillations. Earlier, the Cotentin peninsula was likely to have been an offshore island.

West of the Cotentin peninsula is the Baie du Mont Saint Michel where, apart from Mont St Michel itself (a Hercynian granodiorite boss; Fig. 6.3c) (see Bonnot-Courtois et al., Chap. 5 in Fort and André, Eds., 2014), the area is wholly depositional. The Baie extends from the Sélune estuary at Avranches to the rocky headland beyond Cancale, the Pointe du Grouin. The combination of the highest tidal range in France with a very low seaward gradient of the nearshore zone has resulted in an enormously wide foreshore zone. Between about 3 and 8 km inland from the present (artificially reclaimed) shore is the ancient cliffline dating from the post-glacial sea-level maximum (about 7000 years ago). This zone was previously saltmarsh, but is now reclaimed to form a polder-type landscape. Seawards of the present shoreline is a wide zone of modern saltmarsh, followed by extensive mudflats (important today for oyster cultivation) and sandbars/spits, the whole extending several km from the shoreline.

West of the Baie du Mont St Michel, from Cancale through St Malo and Dinard to St Brieuc, the coast is dominantly an erosional coast with cliffs cut into Hercynian metamorphic rocks. There are local bayhead beaches seaward of old cliff lines, plus the deep ria of the Rance between Dinard and St Malo. The dominantly cliffed coast continues beyond St Brieuc to Île de Bréhat/Paimpol, with small bayhead beaches, but also mudflats in Paimpol Bay, locally protected by offshore rocky islands, and again devoted to oyster cultivation.

The most spectacular part of the north coast is the long section between Paimpol and the most northwesterly point of mainland France, the Pointe de St Mathieu, west of Brest. This is mostly an erosional cliffed coast cut in a variety of Hercynian granitic rocks. Exposed to northwesterly gales, wave action is high because of the long fetch into the Atlantic. There is however, a variety of landform detail, including granite cliffs (e.g. at Pointe de Van, Fig. 6.3a), huge deep rias, especially between Paimpol and Tréguier. Further west, WNW of Lesneven the rias are fault-aligned. There are also small bayhead beaches, plus a few larger sandy bays (Baie de Kernic, Grève de Goulven). There are also locally old cliff lines and offshore rock platforms.

Perhaps the most bizarre part of this coast is the so called "Côte de Granit Rose" (the Pink Granite Coast) northwest of Lannion (see Lageat, Chap. 6 in Fort and André, Eds., 2014). Here, Hercynian granites had been weathered by tropical deep weathering during the Neogene (?) to produce huge corestones, which on later removal of the matrix produced a bizarre boulder-dominated landscape of spectacular granite tors.

The Atlantic Coast The Atlantic coast differs from the Channel coast by the presence of large embayments, for example the Rade de Brest, a huge drowned harbour with an ENE-WSW structural alignment. It is headed by the Aulne ria/estuary, inland of which are incised meanders. To the south is another similar large embayment, the Baie de Douarnenez.

South of these two embayments is the Pointe du Raz, almost the most westerly point of mainland France, with its spectacular granite cliffs. To its south lies the Quiberon coast, initially with massive dunes, then towards Vannes is a complex area including rias, low cliffs, beaches, spits and saltmarshes (south of Lorient). Beyond the Quiberon coast there is an enclosed bay with numerous islands, the Golfe du Morbihan. A visit to this area could profitably be combined with visits to the numerous prehistoric sites at Carnac. From Quiberon the offshore islands might be worth a visit by ferry. There is another complex section of coast between Quiberon and La Baule/St Nazaire at the mouth of the Loire, including the confined estuary of the Vilaine. Inland are the marshlands and the lagoons of La Brière, presumably related to a former position of the coastline? Inland on the Loire estuary the channel is tidal upstream towards Nantes. There is an abandoned channel and reclaimed marshland to the south. Beyond the Loire the coast is mostly depositional, with sandy beaches and reclaimed marshes inland (the Marais Breton). It is possible to drive to the Île de Noirmoutier offshore (sand spits, mudflats). The sandy beaches, in places backed by dunes, continue to Les Sables-d'Olonne, beyond which you would be in the northern part of the Basin of Aquitaine (see Chap. 10).

Highlights: Coastal Sites: Cotentin and Adjacent Areas

The Carentan Estuary (on the east side of the Cotentin peninsula) A visit here could be combined with a visit to one of the 1944, second World War, Normandy landing beaches (Utah Beach, northwest of Carantan). The other Normandy landing beaches (Omaha, Gold, Juno, Sword) are further east. The western side of the Carentan estuary includes small areas of natural non-reclaimed tidal saltmarsh, accessible by the minor road between Carentan and Le Grand Vey. North-West of Grand Vey Point (small sand spits) the coast (Utah Beach) comprises a low sandy beach, backed by low stabilised dunes.

Cap de la Hague/Nez de Jobourg (NW tip of Cherbourg peninsula, Fig. 4.7a). This is a spectacular cliffed coast (at the Nez de Jobourg), with cliffs cut in Hercynian metamorphic rocks. A ridge with elevations 130–180 m along the southern margin of the Jobourg peninsula preserves (Neogene) erosion surfaces which bevel the underlying structures in the Hercynian rocks. Further north, around the Cap de la Hague and east to La Coque, the coast is of lower rocky headlands, locally fronted by sand and shingle beaches. Inland, the low platforms at elevations around 30 m are former wave-cut platforms, dating from higher sea levels probably during the last interglacial or an earlier Mid-Pleistocene interglacial.

Estuaries and the depositional coast on the west side of the Cotentin peninsula between Carteret, Portbail and Lessay (Figs. 4.7b, 6.3b). A series of small estuaries on the west coast of the Cotentin Peninsula exhibits a range of tidal depositional morphologies including: estuary-mouth sandspits; active saltmarshes in sheltered locations; and low aeolian dunes on stable sandspit features.

Coastal Sites: Armorica, North Coast

Baie du Mont Saint Michel (between Carolles, Avranches, and Cancale, Fig. 6.3c) (See also Courtois et al., Chap. 5 in Fort and André, Eds., 2014). The monastery of Mont St Michel, one of the most important tourist attractions in Normandy, is built on a Hercynian-age granodiorite boss on the south side of the Baie du Saint Michel.

In addition to its historical/tourist significance, it has considerable geomorphological interest. The bay has the highest tidal range in France (up to almost 15 m), which when coupled to the very low offshore gradients leads to very hazardous rapid tidal rises. It also means a wide range of coastal depositional environments over a wide area of the bay, including: reclaimed and protected saltmarsh (salinas and oyster beds); natural saltmarshes; creek systems; sandbars; spits; and other depositional features. The high tidal range has been harnessed by the tidal power station on the Rance estuary, inland from Dinard/St Malo.

The "Pink" Granitic Coast (NW of Lannion, between the villages of Trébeurden and Ploumanac'h) (See also Lageat, Chap. 6 in Fort and André, Eds., 2014). Granite tor corestones were produced by chemical deep weathering of the Hercynian granite resulting from the warm humid climates during the Neogene. During the Pleistocene most of the weathered matrix was removed, resulting in the exposure of the corestones (huge granite boulders) in an extraordinary boulder coastal landscape. Some have been given twee names (e.g. The tortoise; Napoleon's hat!). Nevertherless they are spectacular and interesting.

Coastal Sites: Armorica/Brittany, West Coast

Pointe du Raz/Pointe du Van (Almost?) the most westerly points on the French mainland, essentially Brittany's Lands End, Fig. 6.3a. Spectacular faulted granite cliffs form the two headlands. Spurs capped by near-flat erosion surfaces at about 60–75 m (mid-Pleistocene?) are former wavecut platforms. The two headlands separated by Baie des Trépassés are aligned along strike-slip faults. The headlands of Hercynian granites form cliffs with sea stacks.

Quiberon Coast A trip from Lorient/Port Louis along the coast through the Carnac/Quiberon area, then around the Golfe du Morbihan through Vannes to the Sarzeau peninsula, provides an excellent overview of a fascinating coastal area. For those with archaeological interests this also provides an opportunity to visit the Mesolithic stone alignments at Carnac.

This is primarily a drowned coast with local rias, and with evidence of former (mid-Pleistocene?) sea levels preserved in the raised wavecut platforms. These create near-horizontal surfaces cutting across the Hercynian basement rocks, up to about 20 m above modern sea level. The modern coastal features relate to the mid Holocene sea-level rise, which also caused the drowning of the many estuaries and particularly of the Golfe du Morbihan. The coast itself, especially west of Carnac, is marked by Holocene dunes resting on the former wavecut surfaces along the shore, inland of which, especially at Port-Louis, are lagoons fringed by saltmarshes. The Quiberon peninsula, especially on its western coast (the Côte Sauvage), also preserves (mid-Pleistocene?) wavecut surfaces. The drowned Golfe du Morbihan is a huge "inland sea" studded with islands, another legacy from the post-glacial mid-Holocene rise in sea level.

Marais Breton and Île de Noirmoutier There is another area of depositional coast, that of the Vendée between Bourgneuf-en-Retz and St Jean-de-Monts, including the île de Noirmoutier. This area is dominated by reclaimed saltmarshes (Marais Breton and Marais de Challans), on the dominantly depositional Vendée coast. Much of this area is devoted to mussel and oyster cultivation. The offshore Île de Noirmoutier (now accessible by road toll bridge from La Barre-de-Monts) is low lying with sandy beaches.

Inland Sites

The Roc'h Trévezel Area (south of Morlaix). This is an open upland landscape (unusual in Brittany) culmimating in tors developed in late Hercynian granite. It forms the highest point in northern Brittany, 384 m (part of the Monts d'Arrée ridge), and gives extensive views northwards towards Morlaix and the north coast, and towards the east, south and west over the plateau surfaces of northern Brittany. This is one of the few areas in Brittany of open moorland (surrounded by the typical "bocage" landscape of small fields separated by earth bank hedges). Six km to the south is another hilltop viewpoint (Montagne St Michel, 380 m) giving more panoramic views especially over the upland surfaces of Brittany to the west and south. The highest points of both hilltops are tors.

Incised Meanders of the River Aulne (from Châteauneuf-du-Faou downstream especially between Pleyben and Châteaulin, north of Quimper). Here there are classic incised meanders set below plateau surfaces. The plateau surfaces cut across an area of complex Hercynian geology (granites intruded into folded Lower and Upper Palaeozoic metasedimentary rocks). The plateau surfaces are Neogene erosion surfaces into which the Pleistocene valley is incised in a series of incised meanders. The modern river flow here is regulated by a series of weirs, so the modern channel is only partly natural.

Fault Aligned Drainage of the River Oust (WNW of Redon, Fig. 6.2b). The River Oust and adjacent Rivers Arz and Claie are aligned along the major WNW-ESE strike-slip fault systems within the Hercynian granitic rocks of southern Brittany. The lightly incised valley of the Oust (similarly its tributaries the Arz and the Claie) follows the faults. Within the valleys the modern river channels exhibit misfit meanders within lightly incised valley meanders.

The Armorican Loire   (Angers to Ancenis and thence towards Nantes, (see Carcaud and Davodeau, Chap. 7 in Fort and André, Eds., 2014). In contrast with the classic "Loire valley" further east in the southwestern part of the Paris Basin (see Chap. 9), the Breton portion of the Loire (i.e. that downstream of Angers) is actually in a valley! The Loire has essentially a transverse course, cutting across Hercynian structures that link the two parts of the Hercynian massif, Brittany and Vendée. Although I have not come across any recent studies relating to its origin, I suspect it stemmed from superimposition probably during the Miocene. Upstream the river crosses from the Cretaceous rocks of the southwestern part of the Paris Basin onto Precambrian and Lower Palaeozoic rocks of the Armorican Massif. These rocks are affected by WNW-ESE structures (strike-slip faults and low angle thrusts, within folded terrain) across which the Loire valley downstream of Angers is transverse.

The valley floor is somewhat constrained, but the channel belt itself (occupying most of the valley floor) is interesting and includes a braided main channel, stable islands and sub-channels. Immediately upstream of Angers are superb sand-bar braids (see Fig. 9.12c). From Angers through Chalonnes-sur-Loire to about Montjean-sur-Loire the channel (including the braid bars) is more or less natural. From Montjean downstream, through Ancenis effectively to Nantes, berms were constructed in the nineteenth century on the bed of the main channel, to concentrate the flow and maintain a navigable channel. The berms are still there. Though no longer maintained, they still affect the detailed patterns of erosion and deposition within the channel. According to geographers Carcaud and Davodeau (Chap. 7 in Fort and André, Eds., 2014), there is controversy as to whether these should be removed to restore the river towards its natural condition as navigation is no longer important, or whether they should be conserved as part of the human heritage of the Loire.

Chapter 7
The Eastern Uplands from the Ardennes to the Vosges

The north-eastern border of France more or less follows the western limit of a large Hercynian massif that extends eastwards into Belgium, Luxemburg and Germany to form the complex Rhine highlands. It comprises several distinct sub-massifs from the Ardennes in the north, through the Hunsrük to the Saarland in the southeast. South of the Saarland, and also included in this chapter although not strictly a Hercynian massif, are the Vosges du Nord (hills developed in Triassic rocks). South of these, in Alsace, are the Vosges themselves. The Vosges are a Hercynian massif, mirror-imaging the Black Forest massif in southern Germany across the Rhine Rift Valley (see Fig. 7.1 map).

The Geology of the Eastern Uplands

The Ardennes, in northeast France, occupy only a small area within France, but extend into Belgium and Luxembourg. Within France they represent a separate set of Hercynian structures from those of Armorica. The Ardennes have more affinity with the (non-igneous) Hercynian rocks and structures of southwest England. The oldest rocks are Cambrian-age schists and quartzites in the core of an anticline to the north of Charleville-Mézières. These rocks are unconformably overlain by Devonian-age dominantly slates. To the north are Carboniferous rocks, which form the Franco-Belgian coalfield (in northern France, concealed unconformably by younger mostly Cretaceous rocks). The massif as a whole was uplifted during the Cenozoic with uplift continuing into the Quaternary (Demoulin and Hallot 2009). To the South of the Ardennes is a sequence of Mesozoic sedimentary rocks extending into the scarplands of eastern France (see Chap. 9). To the southeast of the Ardennes, extending into Luxemburg, is a lowland belt on Mesozoic (Triassic and Jurassic) rocks which is followed in part in the west by the River Meuse and in the east by the Moselle River. East of this lowland, and wholly in Germany but

© The Author(s), under exclusive license to Springer Nature
Switzerland AG 2025
A. Harvey, *The Geomorphology of French Landscapes*,
https://doi.org/10.1007/978-3-031-68490-6_7

Fig. 7.1 Map of the eastern borders from the Ardennes to the Vosges

extending almost to the French border, is the Hunsrük massif. That area, as are the Ardennes, is formed of lower Devonian sedimentary and low-grade metamorphic rocks. Downfaulted to the southeast of the Hunsrük is the Carboniferous to Permian Saar (coal) basin, again barely reaching the Franco-German border.

South of the Saarland, and extending into Alsace is a hill area, the Vosges du Nord. Though not strictly a Hercynian massif, this area is considered here, to give some spatial continuity. It is formed of Triassic rocks, lower Triassic sandstones, and especially the Middle Triassic limestone (the Muschelkalk), overlain to the west by upper Triassic sandstones and marls. The lower Triassic sandstones continue south where they flank the truly Hercynian rocks and structures of the Vosges them-selves. The central core of the Vosges massif comprises Hercynian gneisses and granites, traversed by major NNE-SSW fault systems, a geology in many ways similar to that of the Morvan, in the northern part of the Massif Central (see Chap. 8).

The Vosges massif lies on the western (French) side of the Rhine rift valley, which developed in the Neogene as part of the crustal accommodation for the Alpine mountain system to the south. The Vosges are more or less a mirror image of the German Black Forest to the east of the downfaulted Rhine valley floor. The Rhine "graben" is bounded by major fault systems, those to its west separating it from the uplifted Vosges du Nord and the Vosges massif itself (Fig. 7.1 map). The fault zone involves a variety of younger rocks, mostly Oligocene to Pliocene sedimentary rocks which also crop out at the head (to the south) of the Rhine rift valley. Incidentally, it is mostly on these downfaulted Cenozoic rocks (whose soils have been mixed by periglacial solifluction) that the Alsace wines are produced. The floor of the Rhine rift is mostly flat, on Late Pleistocene to Holocene fluvial sedi-ments. On the German side of the rift (NE of Freiberg) are Cenozoic volcanic rocks of the Obergergen.

Throughout this area (the Ardennes to the Vosges), the sequence of Palaeozoic rocks was folded then uplifted during the Hercynian orogeny primarily during the Permian period. Any original relief was eroded during the Mesozoic. The whole area has been uplifted during the Cenozoic, uplift which continued into the Quaternary albeit at a reduced rate (Demoulin and Hallot 2009). The present wide-spread plateau-like surfaces were created during the Neogene (for possible origins see Chap. 3). A final phase, deep dissection of these surfaces, took place during the Quaternary (see below).

The Geomorphology of the Eastern Uplands

In the northern part of this area, the Ardennes within France and southern Belgium form a low "upland" plateau with surface elevations ranging between 350 and about 450 m. Deeply incised below this surface is the River Meuse and its tributaries (Figs. 7.2a, b). As elsewhere in France the switch in geomorphic style from plana-tion, by whatever combination of processes, to deep dissection occurred at some stage during the early Pleistocene. This incision in the Meuse valley, and particu-larly its tributary from the east, the Semois, has formed some spectacular, tortuous, incised meanders. By and large the rivers follow the outlines of these palaeo-meanders. Only locally, on the Semois rather than on the Meuse, is there a flood-plain sufficiently wide to allow the development of much smaller "misfit" modern meanders, whose geometry relates to modern rather than Pleistocene (periglacial) runoff conditions (see Chaps. 3 and 4). Also on the Semois, at two locations both just over the border in Belgium, there are cutoff, abandoned palaeo-meander loops (Fig. 7.2b). South and southeast of the Ardennes the geomorphology of the Meuse and Moselle river systems is dealt with later, in the chapter on the eastern scarplands

Fig. 7.2 The Ardennes. (**a**) Les Dames de Meuse, south of Revin: View downstream into the steep undercut slope on the outer bank of the incised meander. Note the narrow floodplain on the inside of the bend (to the right). Note also the bevelled upper surfaces, Neogene (?) erosion surfaces, truncating Hercynian-folded Cambrian sedimentary rocks. (**b**) Landsat Image of the incised meanders of the River Semois [E 4.54′ N 50.00′] (a right-bank tributary of the Meuse), cut into Hercynian-folded Devonian sandstones. Note the limited Holocene floodplain development on the insides of the bends. Note also the abandoned incised meander (centre-right of the image)

(Chap. 9). Mention should be made here of the River Chiers, a right-bank tributary of the Meuse, with its confluence just upstream of Sedan. This river skirts the Franco-Belgian border just south of the Ardennes, and near Montmédy forms impressive incised meanders cut into a Jurassic limestone plateau.

Further south, the Vosges du Nord are drained on their western side by the headwater tributaries of the Saar and its tributary the Blies, both ultimately tributary to the Moselle within Germany. The eastern flank of the Vosges du Nord is drained by small steep streams that feed into the northern part of the Rhine rift. The Vosges du Nord only reach elevations of about 500 m, but further south the Vosges themselves (Fig. 7.1 map) reach elevations of over 1400 m in the Ballons area (Fig. 7.3a, Ballons des Vosges 1424 m). Only 25 km to the east the elevation of the floor of the Rhine rift valley is at only c150 m, resulting in very steep small incised drainages on the eastern face of the Vosges.

Quite clearly the Vosges have been tectonically uplifted by "mid-Tertiary" (Alpine) tectonics and concurrent faulting forming the Rhine rift. The main plateau surfaces in the Vosges are at about 1000 m above the Rhine Rift (those on the Triassic rocks of the Vosges du Nord are at only about 500 m elevation). As elsewhere, there is a question of the age and origin of the plateau surfaces. In the main Vosges, do they represent the stripped sub-Triassic surface? Or are they simply "Tertiary" erosion surfaces? In the northern Vosges, presumably such surfaces are of "Tertiary" erosional origin as they cut across the Triassic rocks. The elevation of the Vosges was sufficient to support a small ice cap during the Pleistocene, which probably reached its maximum during the penultimate glaciation, but was clearly present during the Last (Würm/Weichsel) glaciation. In the Ballons area there are remnant cirques, somewhat degraded, plus other glacial features and moraines. Recent work by Mercier (Chap. 16 in Fort and André, Eds., 2014) has demonstrated the presence of over 40 cirque forms within the highest areas of the Vosges, facing

Fig. 7.3 The Vosges. (**a**) The central Vosges, looking from the Route des Crêtes, into the 'Ballons' area. The skyline is of (Neogene?) erosion surfaces truncating the dominantly granitic Hercynian bedrock, deeply dissected by Pleistocene glacial valleys. (**b**) The Ribeauvillé fault zone on the eastern margin of the Vosges, one of the main source areas of Alsace wines: Shutterstock image 736,226,296 (*copyright: Tanja Midgardson, Shutterstock 736,226,296*)

both eastern and western sides of the main ridges. These fed valley glaciers that extended for 30+ km to the west and for c 10 km down the steeper eastern side. Small corrie lakes are still present in the source areas. There is a range of morainic deposits along the valley floors, plus fluvioglacial terrace forms in the upper Fecht valley (east of Hohneck mountain) mostly of Würm age, dating from a range of deglaciation phases at the end of the last glaciation.

The modern drainage is steep and deeply incised below the plateau surfaces, especially on the eastern side above the Rhine rift (Figs. 7.3b, 7.4 map). On the flat floor of the Rhine rift, west of the Rhine itself the River Ill flows north, parallel to the Rhine. It has an irregular, meandering, locally anastomosing channel. The Rhine previously was a large meandering river, but now is totally managed for navigation. Only in the wholly German sector, north of Lauterberg, is there obvious evidence preserved in floodplain meander scrolls of the previous tortuously meandering channel.

Within the Vosges the streams west of the divide form the headwaters of the Moselle and the Saône drainages. Within the Moselle drainage, the upper Moselle itself above Épinal has a low-sinuosity meandering channel, but upstream of Remiremont former meander cutoffs can be identified in the floodplain. Below Épinal the channel is similar, but there has been more floodplain management. The upper Meuthe, a Moselle tributary, has a single thread meandering channel

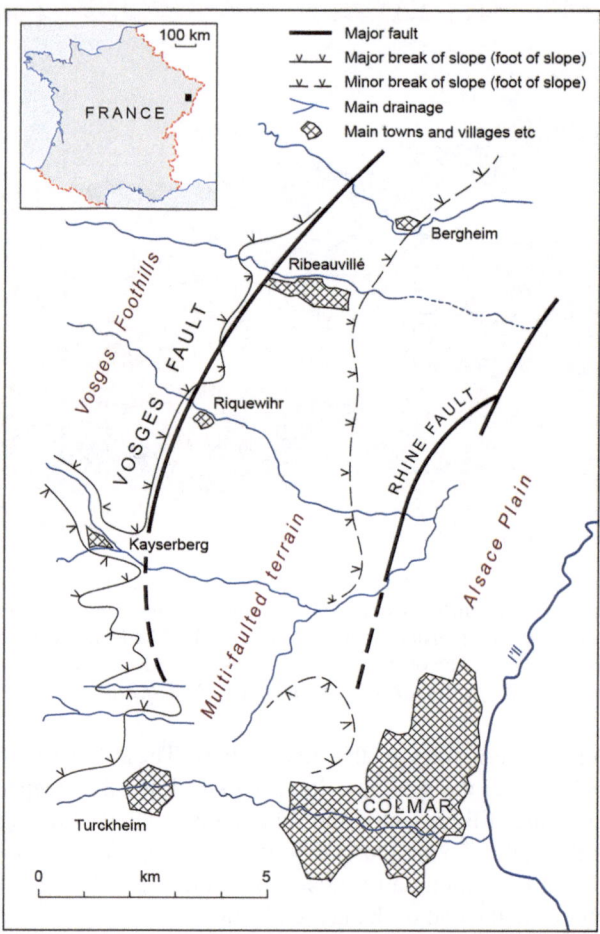

Fig. 7.4 Map of the Ribeauvillé fault zone. The multi-faulted terrain at the foot of the Vosges, on the margins of the Alsace Plain

(incredibly tortuous in a wide floodplain, north of Baccarat). Within the Saône drainage, the upper Saône is tortuous on a wide valley floor, but there is an incised reach near Ameuville. Downstream of Vesoul there are modern misfit meanders within lightly incised valley bends. Further south, the Ognon, a tributary to the Doubs (see Chap. 12), also has a single thread meandering channel, with evidence of numerous cutoffs. Downstream of Rougement, there are big valley bends within which the modern meandering channel of the Ognon is markedly misfit (see Fig. 4.5d).

Highlights: The Ardennes

River Meuse: Incised Meanders (especially from Bogny to Revin) (Fig. 7.2a). The Ardennes plateau is a Neogene erosion surface at an elevation of around 400 m cut across folded Upper Palaeozoic sedimentary rocks. In undisturbed sites it is likely to be mantled by a residual deeply weathered soil, but I do not know of specific locations where such a soil might be seen. During the Neogene the meandering River Meuse was almost certainly superimposed from a cover of "Tertiary" sediments onto the underlying folded Palaeozoic rocks. Then during the Pleistocene the river incised into these rocks, creating the modern spectacular incised meanders. The meanders take a variety of forms ranging from very narrow incised valley floors with almost no floodplain to much more complex forms especially near Givet on the Belgian border. Here there is preservation of erosional stages on the spurs between meanders, and of depositional terraces on the valley floor. In places the valley walls are steep, especially on the outsides of bends (eg. Les Dames de Meuse, between Laifour and Revin, Fig. 7.2a). Within the incised valley there is very little alluvial valley floor; what there is occurs mostly on the insides of meander bends. The modern channel occupies most of the valley floor, and is mostly single-thread (with occasional alluvial islands). Flow levels are maintained artificially by weirs, which are by-passed locally by short, navigation canals and locks.

River Semoy/Semois: Incised Meanders with Modern Misfit Channel Meanders (Fig. 7.2b). Even more spectacular and varied than those on the Meuse are the incised meanders on the Semoy, an east-bank tributary of the Meuse. Upstream from its confluence with the Meuse, the River Semoy/Semois cuts a similar incised valley into the plateau. The plateau itself is cut across the Hercynian folded Palaeozoic sedimentary rocks. There are similar incised meanders to those on the Meuse, but on the Semoy they are exceptionally well developed. There are excellent viewpoints above the valley from the Meuse confluence at Monthermé (panoramic viewpoints above the village) to Les Hautes-Rivières (on the Belgian border). The incised meandering geomorphology includes the modern misfit meandering channel within the large valley meanders. It would be worth continuing upstream into Belgium, at least to Vresse, to see similar valley/channel relationships, but also two cutoff valley meanders (south and southeast of Vresse).

Highlights: The Vosges

The Ballons des Vosges/Hoeneck Areas   (Fig. 7.3a), (See also: Mercier Chap. 16, in Fort and André, Eds., 2014). Southeast of Gérardmer and centred on Hohneck (1363 m), is the highest part of the Vosges massif including the Grand Ballon (1424 m) in the east and the Ballon d'Alsace (1250 m) in the south. This is the core of the area glaciated during the Late Pleistocene. The summit areas and the upland plateau surfaces are remnant erosion surfaces into which the valley systems are incised. There has been (unresolved) debate in the research literature on the origin and age of these surfaces. Following Alpine-age uplift and rifting, they clearly pre-date Pleistocene incision. During the Mid-Pleistocene, then again during the Late-Pleistocene, glaciers formed in these valleys. Very little evidence remains of the Mid-Pleistocene glaciers, but there is plenty of evidence, both erosional and depo-sitional, for the Late Pleistocene glaciers. The steep glaciers flowing east from the divide area extended about 10 km from the divide area, whereas those to the west extended for up to about 30 km or more. Within the most intensely glaciated area are numerous cirques feeding into U-shaped trough valleys. Further down the val-leys (eg. the Fecht valley southeast of Hohneck) are moraines. Throughout the area Latest Pleistocene periglacial features (screes, scree cones etc.) are important.

The Ribeauvillé Fault Zone, the Margins of the Rhine Rift   (Figs. 7.3b, 7.4 map), (see also Carozza, Chap. 14 in Fort and André, Eds., 2014). This is a very distinctive landscape, associated with the fault zone bounding the eastern margin of the Vosges. It is an attractive landscape that coincides with the main wine growing area of Alsace. The area extends north to south through almost 90 km from the latitude of Strasbourg to that of Mulhouse. Perhaps one of the most characteristic and interest-ing parts of the area, as well as being an attractive area itself, is the Ribeauvillé/ Riquewihr area northwest of Colmar. The area is bounded on its western side by the main Vosges margin fault, separating the Hercynian granitic terrain to the west from the multiply-downfaulted "Tertiary" rift-valley fill to the east. The surficial deposits, derived by Pleistocene solifluction processes, are a mixture of granitic material derived from the Vosges, and calcareous materials derived from the downfaulted Cenozoic sedimentary rocks. Morphologically these deposits form colluvial slopes below the steeper Vosges margins but above low-angle fans that are inset within the colluvial slopes. Further to the east the slopes grade into the wetlands of the Ill val-ley drainage. The colluvial slopes are those devoted to vine cultivation.

Chapter 8
The Massif Central (Limousin, Auvergne, Causses)

Introduction

The Massif Central (Fig. 8.1 map), although one of the Hercynian massifs, is much more varied with much more exciting geology and geomorphology than the massifs dealt with so far (Armorica, The Ardennes, The Vosges).

As with the other massifs, in the Massif Central the core is of Hercynian metamorphic and granitic rocks, in structures relating to the southern (Gondwana) flank of the Hercynian collision zone. A major N-S fault system separates the dominantly granite terrain of Limousin to the west from the more complex terrain of Auvergne to the east. Much younger rocks and structures are incorporated into the massif, especially in the southeast. On the southern flanks of the massif are the Cévennes, ranges formed of lower Palaeozoic rocks deformed by Palaogene (Pyrenean) east-west fold structures. Trapped between these areas and the southern part of the massif itself are the uplifted plateaux of Jurassic limestones of the Grands Causses/Gorges du Tarn area (see below). The whole of the Massif Central was uplifted by "mid-Tertiary" (Alpine) tectonics. The present maximum elevation of Auvergne (up to 1855 m) gives an indication of the amount of Cenozoic uplift. Compare this figure even with the figure of 1400 m + for the Vosges. Furthermore it is markedly more than the approximately 450 m for both the Ardennes and Armorica. A major fault system developed along the eastern margin, separating the massif from the downfaulted Rhône/Saône corridor and in the south from the Cretaceous limestone Ardèche plateau. Similarly, the downfaulted Limagne/Allier and Roanne/Loire depressions were formed in the north (in a way mirroring the Rhine rift valley). The final geological phase, also related to post-Alpine conditions, was the Neogene to Quaternary volcanicity in Auvergne. The Quaternary also had two other impacts. (i) Several small ice caps were formed in the Sancy, Cantal, Aubrac and Forez areas. (ii) The main drainages were incised into bedrock, resulting not only in incised meanders on most of the drainages, particularly the western drainages, but also in

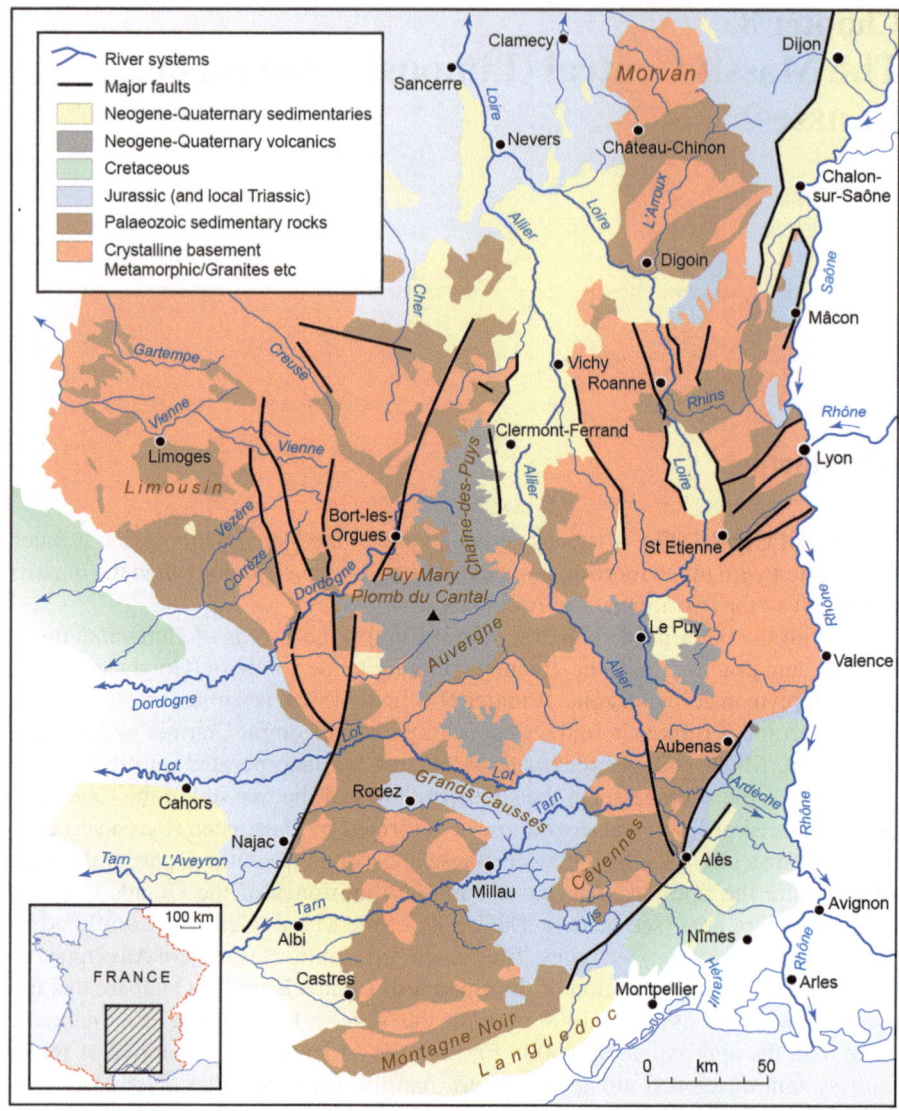

Fig. 8.1 Map of the Massif Central

several spectacular gorges and canyons, especially in the south, notably the Gorges du Tarn. (For locations see Fig. 8.1 map).

The Geology of the Massif Central

The central core areas of the Massif Central comprise metamorphic rocks and granites, deformed and emplaced during the Upper Palaeozoic, Hercynian orogeny. These rocks dominate Limousin, the area to the west of the major NNE-SSW fault that separates Limousin from Auvergne. Auvergne is more complex. Its basement comprises similar granitic and metamorphic rocks to those in Limousin, but it includes a much wider range of post-Hercynian rocks and structures. North of Auvergne, in western Burgundy, is the most northerly part of the Massif, the Morvan (Fig. 8.1 map). This area, dominantly of granites, has affinities with the Vosges and is dominated by an ENE-WSW structural alignment, but with incorporated Carboniferous volcanic and sedimentary rocks. Its western, northern and eastern margins are flanked by the Jurassic limestones of northern Burgundy.

South of the Morvan the eastern margin of the Massif Central stands above the downfaulted Rhône/Saône corridor (see Chap. 12). Related to these ("Mid-Tertiary" Alpine-age) bounding faults are a series of NE-SW faults affecting the whole of the eastern margin of the massif. The first major fault in this series bounds the Morvan to its south, beyond which lies the Upper Carboniferous (Coal Measures) coal basin including the iron ores of Le Creusot. Beyond the southern bounding fault of that basin is a sliver of granite, culminating in one of the highest points of southern Burgundy (at about 600 m), Mont-St-Vincent. Further south are the Hercynian granites of Charollais, then more Carboniferous volcanic rocks and Hercynian granites in Beaujolais. Further south still, southwest of Lyon, more NE-SW (Alpine-age) faults bound another Coal Measures basin, the Saint-Étienne basin (Fig. 8.1 map). South of that basin is a large area of granitic rocks, forming the Monts d'Ardèche. Beyond this a major NE-SW fault, running through Privas and Aubenas to Alès. It truncates the Massif Central and the eastern flank of the northern Cévennes, separating them from the Cretaceous limestone of the Ardèche Plateau to the east.

The central part of the Massif Central through Auvergne is equally complex. In the north are two downfaulted graben structures (Fig. 8.1 map) similar in many ways to the Rhine rift. These are the Limagne/Allier and Roanne/Loire basins. They are early Alpine in age (Palaeogene downfaulted) and floored by Oligocene sediments. Between the two are the granites of the Livradois/Forez area. To the west and south of the Limagne/Allier depression are the Neogene-Quaternary volcanics of the Chaîne des Puys (Fig. 8.2 map). Further south is the huge Miocene volcanic edifice of the Monts du Cantal/Puy Mary (Fig. 8.2 map). To the southeast is the Devès lava plateau, the Velay and the Mont Mézenc Neogene-Quaternary volcanic centres (Fig. 8.2 map, see also Fig. 3.3a–c).

South of the volcanic areas are more Hercynian granites forming the Margeride range, which marks the southernmost part of the Massif Central proper. However in

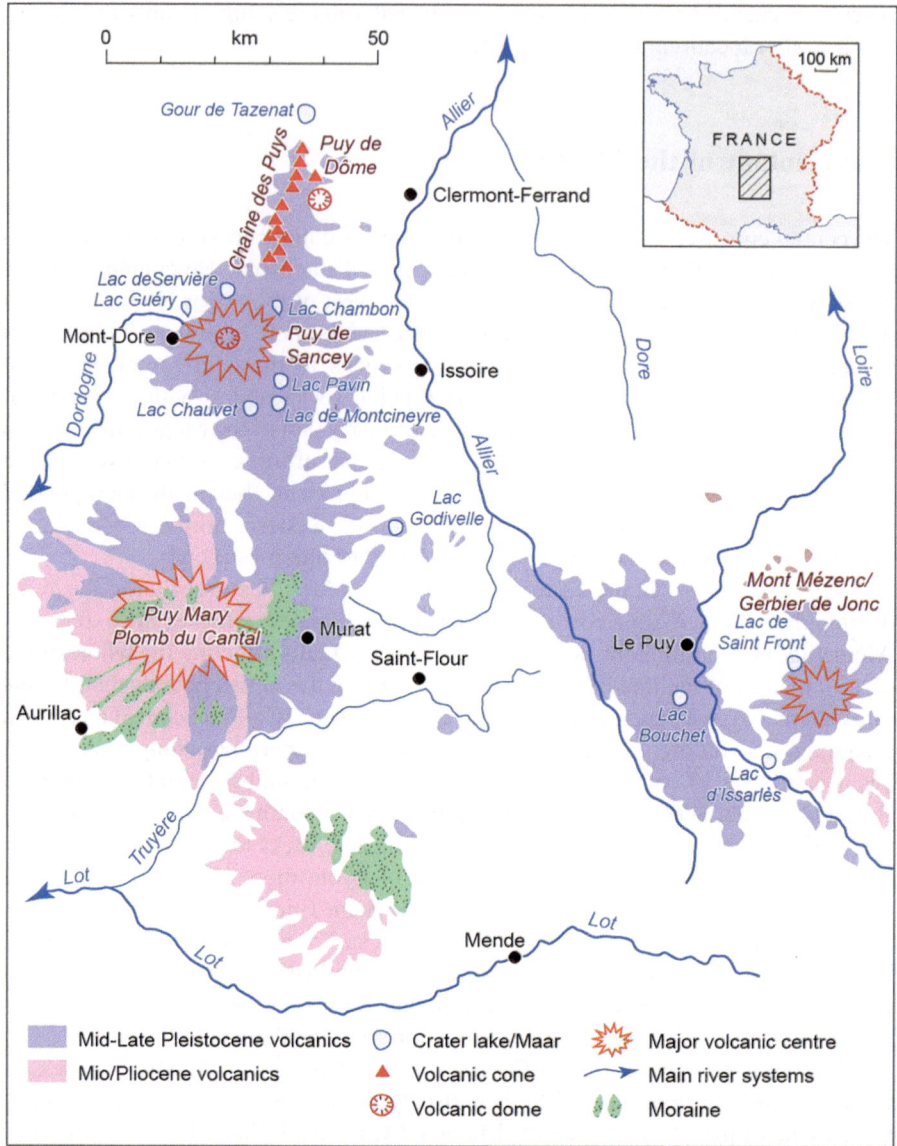

Fig. 8.2 Map of the main Neogene-Quaternary volcanic terrain of Auvergne (Also shown is the main Pleistocene morainic terrain)

a broad sense the high terrain continues into the uplifted karstic plateau of Jurassic limestones of the Grands Causses. These are bounded to the east by the granites and metamorphic rocks of the Cévennes and to the south by similar rocks within the Monts de Lacaune and the Montagne Noire. These latter are folded east-west, along a Pyrenean trend.

Beyond the major fault system that bounds Auvergne to the west, the geology of Limousin (the western part of the Massif Central) is relatively simple. It is dominated by Hercynian granites. These were intruded into a range of country rocks, including Late Precambrian and Lower Palaeozoic country rock. The western margin is mostly fault bounded from the Jurassic rocks that form the margins of the Aquitaine basin.

The Geomorphology of the Massif Central

The geology of the Massif Central, particularly of Auvergne, is exceptionally diverse. The same is true, perhaps even more so, of the geomorphology. In the north (the Morvan) and in the west (Limousin) the landscape is duller, with both areas dominated by ancient (Neogene?) erosion surfaces cutting across dominantly granitic Hercynian bedrock.

The Morvan and Related Areas The Morvan reaches maximum elevations of c900 m in the south of the area but only c600 m in the north. As with the other Hercynian massifs the upland surfaces are (Neogene?) erosion surfaces which truncate the underlying Hercynian structures. The Morvan is drained to the north and northwest by Seine tributaries, the Serein and the Yonne, and to the south and southwest by Loire tributaries, the Arroux and the Aron. These rivers mostly have courses within incised valley meanders, locally showing modern misfit meandering channels within the valley meanders (see Fig. 3.5 map). The Morvan is bordered to the west, north and east by gentler country on Lower Jurassic rocks, and to the south by a fault separating it from the Le Creusot coal basin.

South of the Le Creusot basin are a series of NE-SW fault-aligned ridges, after which are the granitic hills of Beaujolais. Beyond Beaujolais into Lyonnais the terrain is again fragmented by NE-SW faults into ridges and basins, including the St Étienne coal basin, southwest of Lyon. Most of the drainage throughout this area is fault-aligned, flowing east into the Saône or the Rhône.

Limousin To the west of Auvergne is Limousin, defined here as that area on Hercynian rocks and structures to the west of the NNE-SSW fault which runs through the centre of the Massif Central. Limousin is bounded to the north and west by the Jurassic rocks of the southern part of the Loire valley and of the eastern part of the Aquitaine basin, which rest unconformably on or are faulted below the Hercynian rocks. The divides are essentially plateau-like erosion surfaces of uncertain age, but, on the basis of soil maturity and depth of weathering on the underlying granite, are clearly pre-Pleistocene. Elevation of the plateau ranges from 900 m + in the east near the bounding fault to between about 250 and 500 m on the western margin (north to south), although there are local higher points. The Limousin plateau does stand much above the surrounding terrain, which ranges in elevation between about 200 m in the north to about 400 m in the south.

In the north, the Limousin area is drained by tributaries of the Loire system (Fig. 8.3 map). East to west the main such rivers are: the Sioule (which crosses the Limousin/Auvergne bounding fault to join the Allier); then the Cher; the Indre; the Creuse and the Vienne (all tributaries of the Loire). The upper Creuse valley is aligned along a major fault system within the Hercynian granitic rocks of eastern Limousin. The Vienne has an interesting course, initially flowing west (through Limoges) but then turns abruptly north, to more or less follow the western bounding faults of the Massif. It appears that the upper westward-flowing course had once continued further westwards to feed into the Charente, only to be captured by the northward- flowing segment of the Vienne, working back along the fault zone (see Chap. 9). In researching this material, I have not been able to track down any French publication dealing with the capture.

The bulk of the Limousin area is drained westwards by Atlantic drainage towards the Aquitaine basin and includes the headwaters of the following rivers: the upper Charente; the Dronne (a north-bank tributary to the Isle); the Isle itself; the Auvézère (a south bank tributary to the Isle); the Dordogne system, including its north-bank tributary the Vézère and its tributary the Corrèze; and the Cère, a south bank tributary to the Dordogne. Both the Corrèze and the Dordogne rise in Auvergne and cross the Limousin bounding fault.

The river valleys themselves form the most impressive landforms of the Limousin region. Within the Limousin massif itself they tend to be deeply incised, whereas in their lower courses either towards the Loire, or in the Aquitaine basin, the rivers tend to have alluvial channels within incised meanders.

The Auvergne Throughout the Massif Central most of the main rivers are deeply incised (Fig. 8.3 map, see also Figs. 1.2b, 3.5 map, 4.5c) within incised meandering valleys, gorges or canyons. This is indicative of ongoing uplift of the Massif Central during the Quaternary. Exceptions are the upper Allier and upper Loire which drain the downfaulted Limagne/Allier and Roanne/Loire basins respectively. In their uppermost reaches even the Loire and the Allier are deeply incised, the upper Loire in the Le Puy area particularly showing an interesting incisional history in relation to the volcanic sequence (see below). On the eastern margin of the Massif Central the right-bank tributaries of the Rhône are steep incised streams, but south of Privas the massif is separated from the Rhône valley by a downfaulted block of Cretaceous limestone terrain, the Ardèche Plateau (see Chap. 14) into which the rivers are deeply incised.

In the west of Auvergne are the headwaters of the Dordogne and Lot systems (the Auvézère and the Upper Dodogne itself, the upper Lot and its tributary, the Truyère). Farther south are the headwaters of the Tarn system including the Aveyron, and its tributary the Viaur. All these rivers have spectacular incised valleys, in many cases showing well developed incised meanders. Elsewhere there are gorge and canyon reaches. Canyons are particularly well developed in the Gorges du Tarn area cut into the Jurassic limestones of the Causses plateau (see below).

The modern river channels include many bedrock-controlled reaches (see Fig. 4.4a) especially within the canyons and gorges. However, the majority of the

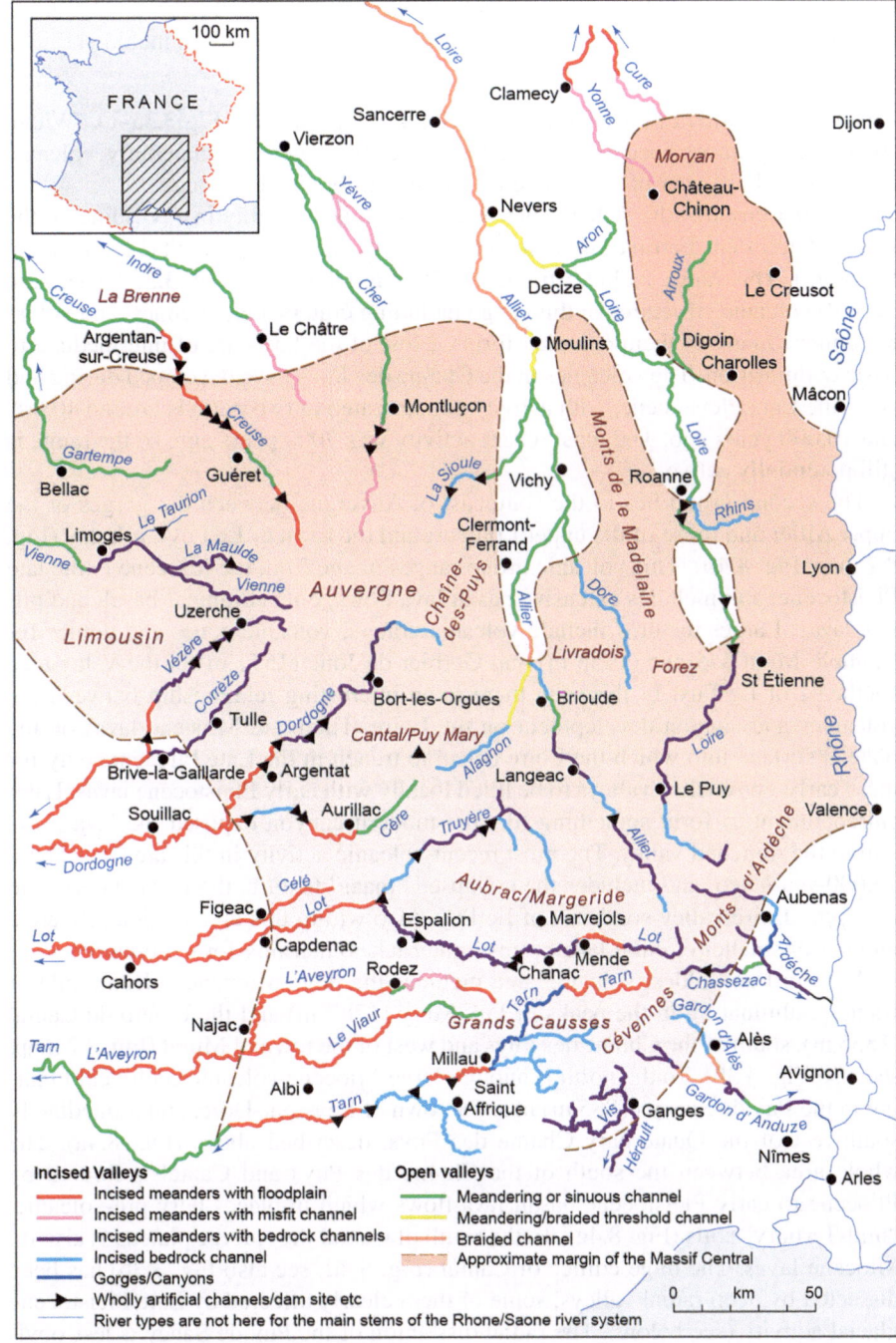

Fig. 8.3 Map of the main drainage network and the geomorphology of the Massif Central

modern river channels are alluvial channels including single-thread mostly mean-dering channels (see Fig. 4.5a), but where sediment supply is high include braided channels.

Auvergne, Volcanic Landforms (Fig. 8.4, See also Fig. 8.2 map, Fig. 3.3a–c). Within Auvergne the most spectacular landforms are the Neogene-Quaternary volcanic landforms. They occur in three more-or-less age-differentiated groups.

The first group is the "Chaîne des Puys" which lies roughly in a N-S line, to the west of Clermont- Ferrand, west of and parallel to the Limagne fault (Fig. 8.2 map, Fig. 8.4a). This range includes the iconic Puy de Dôme (see Fig. 3.3c). There are over 70 volcanic structures in this range including craters, cinder cones, and volca-nic domes, many of them complex forms. Most of the lavas are of intermediate to basic composition. The volcanics in the Chaîne des Puys have developed during and since the Late Pleistocene, with activity concentrated in two periods around 40,000 and 10,000 years ago. The most recent activity was 7000 years ago, so the range is still potentially active.

The second group lies to the southeast of Auvergne between the gorges of the upper Allier and those of the upper Loire around the town of Le Puy-en-Velay (Fig. 8.2 map, Fig. 8.4b). This volcanic group ranges in age from the Miocene to the late Pleistocene, and includes extensive basalt lava flows, both columnar basalt and pil-low lavas. Larger features include volcanic craters, volcanic cones and domes for example Mont Mézenc (1753 m) and Gerbier de Jonc (1551 m) in the Velay area southeast of Le Puy. In this area there is an interesting relationship between the volcanics and canyon development on the Loire. There are Miocene lavas on the plateau surface into which the Loire began to trench in the Late Pliocene, only for these early-entrenched valleys to be filled locally with early Pleistocene lavas. Later entrenchment to form something like the modern canyon exposed the lava flows within the trenched valley. The most recent volcanic activity in this area was about 75,000 years ago, and includes the collapsed "maar" feature, the Lac d'Issarlès, in the upper Loire valley southeast of Le Puy. Also within this area particularly are a number of smaller circular lake basins (crater lakes) marking former craters.

The third and oldest volcanic area includes the huge Miocene volcanic pile of Cantal, culminating in the peaks of Puy Mary (1787 m) and the Plomb du Cantal (1855 m), south of the Chaîne des Puys and west of the town of Murat (Fig. 8.2 map, see also Fig. 3.3b). To the north is another large Miocene volcanic centre culminat-ing in the Puy de Sancy (1885 m) near the town of Le Mont-Dore, and immediately southwest of the Quaternary Chaîne des Puys, described above (Fig. 8.4a). The whole area between the south of the Chaîne des Puys and Cantal is floored by Pliocene to early Pleistocene basalt lava flows which in places bury pre-volcanic, "mid-Tertiary" soils (Fig. 8.4c). To the south of Cantal the Aubrac plateau is also on Miocene lavas. The huge edifice of Cantal (Fig. 8.4d; see also Fig. 3.3b) has been dissected by deep radial valleys, some of them clearly affected by Late Pleistocene glacial activity (see below). The radial dissection of the Puy de Sancy is less obvi-ous, but here there is also clear evidence of Pleistocene glaciation (see below).

Auvergne, Glaciation of the Volcanic Area A second aspect of the geomorphol-ogy of the higher parts of Auvergne that distinguishes Auvergne from Limousin and

Fig. 8.4 Volcanic terrain of Auvergne. (**a**) Plio-Pleistocene volcanic terrain: The Chaîne des Puys, south of Puy de Dôme (see also Fig. 3.3c). (**b**) The Late Neogene to Quaternary volcanic neck at Le Puy, on the upper Loire: Shutterstock image 1,379,282,003 (*copyright: stefano cellai, Shutterstock 1,379,282,003*). (**c**) Late Neogene-Quaternary basalt lavas burying a pre-volcanic soil (a palaeosol), near Salers, Monts de Cantal, southern Auvergne (see also Fig. 3.3b). (**d**) The enormous late Miocene Cantal volcanic complex (see also Fig. 3.3a, b) includes rugged and spectacular terrain, deeply dissected by a radial drainage network, which itself was modified by Pleistocene glaciations. View looking up-valley (a former glacial trough) towards the summit of Puy Mary (see also Fig. 3.3b)

the Morvan, is Pleistocene glaciation. Neither Limousin nor the Morvan show any sign of Pleistocene glacial ice, but there were glaciers in parts of Auvergne. There is clear evidence of glacial erosion in both Sancy and Cantal in the form of cirques at the valley heads, and glacially eroded troughs (Fig. 8.4d). Presumably there are moraines lower down the valleys, but I do not know any details. Pleistocene glaciation has been reported in the Aubrac to the south of Cantal, and also in Velay, and Forez to the north of Velay. I do not know these areas well, nor can I find detailed accounts in the research literature of glaciation in these areas. Within the Massif Central it is only in Forez (on Hercynian granitic basement rocks) that there is a suggestion in the terrain of the presence of glacial cirques, particularly on the north slope of Le Puy Gros, to the northeast of Ambert. It is possible that the southwest slope of Aubrac has been scoured by glacial erosion, but I am not convinced. In Velay, I cannot find detailed evidence of glaciation, but there are several mentions of the possibility of intense periglacial activity (blockfields), which in itself suggests an absence of glaciation. The glaciation identified in Sancy and Cantal is ascribed to the last glaciation. I suspect the glacial extent during the penultimate glaciation would have been greater, but again, I can find no details in the literature.

Southern Auvergne (the Cévennes; the Montagne Noire; the Karst Landscapes of the Causses Plateau) To the south of the Margeride granite and the Aubrec volcanics lies a complex terrain: the Causses karstic plateau on Jurassic limestones, bounded by three upland blocks on igneous and metamorphic rocks. These blocks are: the Plateau du Lévézou to the west; the Cévennes to the east; and the Montagne Noire group to the south. The latter two, particularly the Montagne Noire group, have been affected by west to east Pyrenean structures of Late Cretaceous-Paleocene age. The Montagne Noire group, including the Monts de L'Espinouse and the Monts de Laucune are essentially a detached set of Pyrenean structures separated from the Pyrenean frontal ranges by the Carcassonne gate on Cenozoic rocks (see Chap. 10).

The Causses Plateau is one of the most significant and spectacular karst landscapes in Europe (Fig. 8.5 map, Fig. 8.6). Karst features are well developed at every scale. On bare rock faces detailed solutional features are evident (Fig. 8.6a). On a larger scale solutional sinkholes and larger depressions (uvula) are present on the plateau surfaces. Much of the drainage is underground, and sinkholes and resurgences are common. However, the most spectacular and impressive forms are the deeply incised gorges and canyons of which two, the Gorges du Tarn (Fig. 8.6b), and the gorges of the Vis are the most significant. The Vis is a headstream of the Hérault, which ultimately drains to the Mediterranean rather than the Atlantic. It has been estimated that the Causses Plateau has been uplifted by up to 1000 m over the last two million years. The response of the Pleistocene river systems has been deep incision forming major incised meanders around the margins of the plateau, but deep rock-cut gorges in the plateau centre, particularly the Gorges du Tarn (see above) upstream of Millau and the incised meanders of the gorges of the Vis, including the Cirque de Navacelles (Fig. 8.6c), a UNESCO "World Heritage Site". It is not an easy place to reach! (25 km by minor roads northeast of Lodève). This is not a

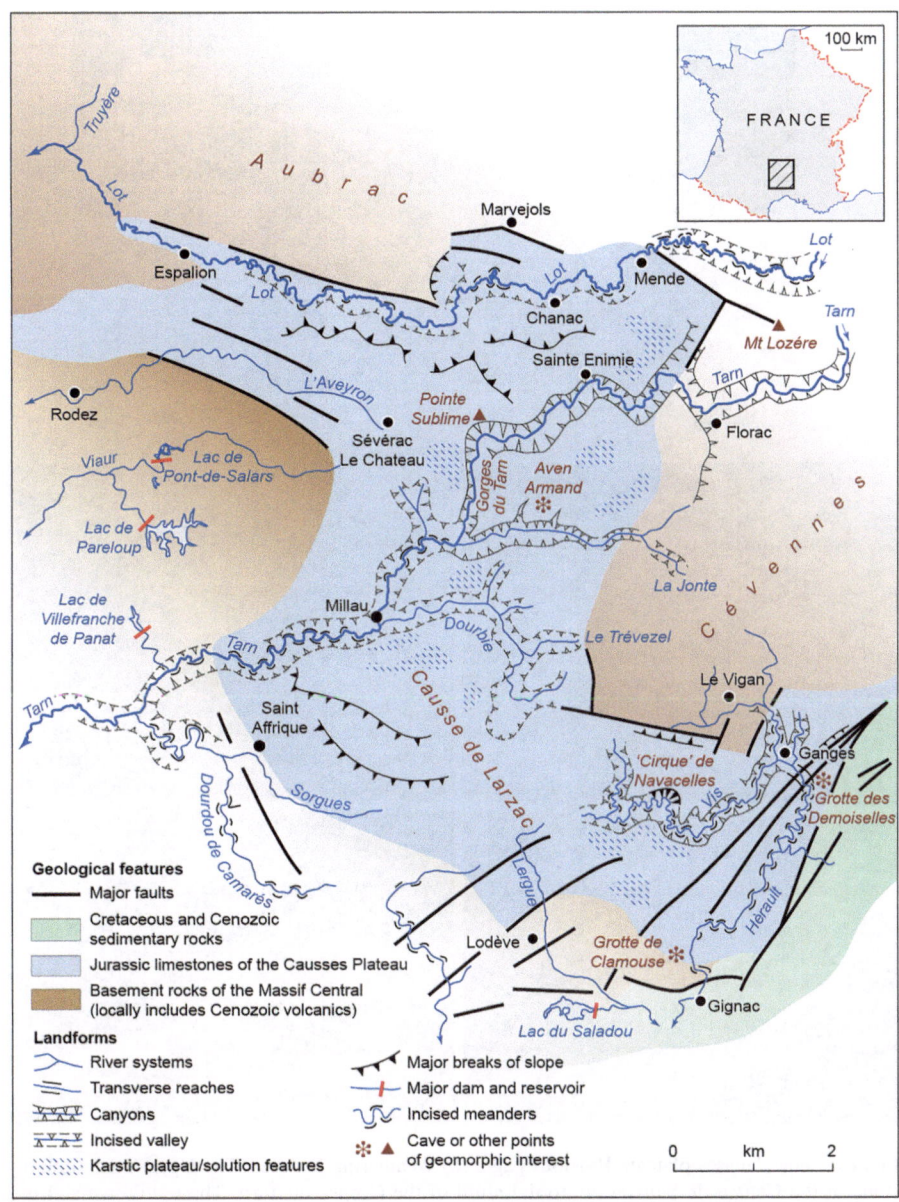

Fig. 8.5 Map of the Causses plateau, karst area in the southern Massif Central

"cirque" in the glacial sense but an amphitheatre created by a meander cutoff. The River Vis originates in the Cévennes to the northeast of this part of the Causses Plateau. Then for its initial course, which is incised into the Causses Plateau, the valley floor is dry. The river flows underground, but emerges as springs downstream

Fig. 8.6 The Causses plateau. Previous page: (**a**) Weathering features: The Upper Jurassic lime-stone of the Causse de Sauveterre, to the north of the Gorges du Tarn. The red/brown soil is a remnant of the highly oxidised Neogene 'Terra Rossa' palaeosol. This page: (**b**) Gorges du Tarn: Detailed view of the bottom of the canyon. With its largely bedrock-controlled river channel there is virtually no space for sediment storage. (**c**) The Cirque de Navacelles: Not a cirque in the glacial sense but the course of an incised meander of the Vis River (a tributary of the Mediterranean Hérault River). The incised meander has been cut off by subsurface karstic flow

of Vissec, upstream of Navacelles. The meander loop itself at Navacelles was occupied by the river during the last glacial, but cutoff occurred in the late Pleistocene or early Holocene, by which time the river was underground. Cutoff was followed in mid-Holocene by the formation of thick travertine deposits on the valley floor, extending several km downstream. These have since been incised by the modern stream, creating a waterfall at the location of the cutoff meander.

Highlights: The Morvan and Related Areas in the Northern Massif Central

The Morvan This is a large and rather diffuse area for which it is difficult to make recommendations for specific site visits that highlight the major features of the geomorphology. It is an upland area of Hercynian granites whose summit areas are bevelled by a set of (Neogene?) erosion surfaces. The Morvan is drained to the northwest by the headstreams of the Yonne including the Cure, to the northeast by those of the Armançon (both Seine headwaters: see Highlights Chap. 9), and to the south by the Aron and the Arroux, both Loire tributaries.

Mont St Vincent This is a hilltop-village viewpoint over the eastern margin of the Massif Central between Saône and Loire drainages (see Chap. 12). The granites of the Morvan are truncated to the south by a NE-SW aligned fault, one of a series of similar faults more or less along the eastern margin of the Massif Central. The first of the faults brings down the Carboniferous rocks of the Montceau-les-Mines/Le Creusot coalfield basin, south of which are the granites of the Mont-St Vincent ridge. Mont-St Vincent itself provides an excellent viewpoint over the dissected terrain of the eastern margin of the Massif Central, drained to the east by the Guy and ultimately to the Saône, and to the west by headwaters of the Bourbince and ultimately to the Loire.

Highlights: Limousin and the Western Massif Central

Limousin As with the Morvan, Limousin is also an area for which it is difficult to identify specific sites as geomorphic highlights. As with the Morvan, the geology is dominantly Hercynian granite, bevelled by Neogene (?) erosion surfaces (the Millevaches Plateau—here "vaches" meaning ponds not cows!). The Pleistocene drainage network is more or less incised below the plateau surfaces. There is one important site that relates to Pleistocene drainage re-alignment, Exideuil on the Vienne.

Exideuil, Capture of the Proto-Charante by the Vienne Just downstream of
Exideuil the River Vienne, which had hitherto been flowing WNW in a broad open
valley, turns abruptly north (Fig. 8.3 map), now flowing in an almost straight course
in a more restricted valley. Continuing westwards from Exideuil is a broad open gap
within which are the modern diminutive headwaters of the River Charante.

Highlights: Auvergne

In the Auvergne we are spoilt for choice in the selection of highlights: Neogene-
Quaternary volcanics, Pleistocene glacial forms, deeply incised canyons.

La Chaîne des Puys (Fig. 8.4a, see also Fig. 3.3c) (see also Boivin and Thouret,
Chap. 9 in Fort and André, Eds., 2014): This zone comprises a set of Late Quaternary
volcanic centres on the Hercynian granitic plateau to the west of the city of Clermont-
Ferrand. The volcanics lie to the west of the Limagne fault, the Neogene fault that
defines the west of the Limagne rift basin. This is the youngest set of volcanic rocks
in France, the latest of a series dating back to the Neogene. La Chaîne des Puys
comprises a suite of discreet volcanic centres with ages ranging from the mid to late
Pleistocene (c90 ka?) to mid Holocene (c7 ka). It forms a series of discrete volcanic
centres with morphologies of four types: simple and compound scoria cones; classic
crater-topped morphologies (composed of intermediate to basic lavas), from which
lava flows emanated; dome-like forms (composed of acid lavas); and explosive/
subsident forms ("marre"). To the west of the volcanic field the River Sioule is
incised deeply into the plateau surface that caps the Hercynian granite. The incised
channel is, as in many such features, characterised by incised meanders, albeit with
some now drowned by a reservoir. Significantly, late Pleistocene lava flows derived
from the Chaine des Puys volcanoes flowed into the Sioule gorges.
     To the south of the Chaîne des Puys is the Mont Dore/Puy de Sancy (1885 m)
complex, an older (early-mid Pleistocene?) volcanic centre, which includes basalt
lava flows (Fig. 3.3a), a volcanic centre now deeply dissected in a more or less radial
pattern, together with a number of crater lakes. An indication of the scale of Neogene
uplift of the central part of the Massif Central is given by the depth of dissection of
all the major drainages, many in deep canyons. Le Mont Dore is bounded to the
west by the canyons of the upper Dordogne. In the Bort-les-Orgues area (more
Neogene volcanic rocks), the incised Dordogne is aligned along the major N-S
Auvergne/Limousin bounding fault that divides the whole of the Massif Central into
two. To the south is the incised River Rhue, a tributary of the Dordogne, and to the
east are incised tributaries of the Allier.

The Plomb du Cantal Neogene Volcanic Centre (Fig. 8.4c, d, see also Fig. 3.3b):
South of the Chaîne des Puys is an enormous spectacular volcanic edifice centred on
the Plomb du Cantal (1855 m), a huge Neogene to Quaternary volcano. The eleva-

tion of the plateau on the Hercynian granitic basement here is around 1000 m, which gives some idea of the scale of the volcano. As with the Mont Dore area further north, some idea of the overall effects of the Neogene tectonic uplift of the Hercynian basement is given by the presence of deeply incised canyons affecting the main drainages on all sides of the Cantal volcano: the Dordogne in the west; the Truyère in the south (with several dams and associated artificial within-canyon lakes); and the Allier headwaters in the northeast.

The lavas building the volcano date from the late Miocene into the Pliocene, with extensive early to mid-Pleistocene basaltic lavas extruded over the lower flanks of the volcano. Once eruption ceased, dissection of the volcano began, with the development of radial drainage in all directions away from the summit areas (Fig. 8.4d, see also Fig. 3.3b). This pattern was accentuated during the Mid- and Late Pleistocene by glaciation. There undoubtedly was glaciation during the penultimate glaciation, but the field evidence is less clear than for the last glaciation. Last-Glacial erosional forms include cirques at the heads of most of the valley systems, and deeply scoured U-shaped main valleys, accentuating the radial nature of the drainage. Lower down the valley systems morainic deposition dominates.

Le Puy Area, Upper Loire Valley and the Volcanics of the Velay (Fig. 8.4b, see also Defive and Poiraud, Chap. 10 in Fort and André, Eds., 2014). A third Neogene-Quaternary volcanic area occurs in eastern Auvergne, in the uppermost Loire valley, the Devès plateau near Le Puy, extending into the Mont Mézenc area within the Velay region to the east of Le Puy. The pre-volcanic terrain of this area involves contrasts between the west and the east. The western part is characterised by low gradient plateau surfaces cut across the underlying Hercynian granitic rocks, later mostly buried by Neogene lavas. In contrast, the eastern part is more deeply dissected in response to widespread incision by the headwaters of the Ardèche (Rhône drainage) following Neogene uplift along the eastern marginal faults of the Massif Central.

There were two phases of volcanic activity in this part of the Massif Central. The first, during the late Miocene, created the lava plateau of the Velay, together with a series of intrusions, domes and cones. The eroded remnants of these form the higher peaks of the area, Mont Mézenc (1753 m) and Le Gerbier de Jonc (1551 m). Springs on the southwest side of the mountain (Le Gerbier de Jonc) incidentally form the source of the Loire. The southeastern side of the mountain is drained by headwaters of the Ardèche system. A late phase in this volcanic activity saw the development, through explosive activity then subsidence, of a number of maare (crater lakes) forming, among others, the Lac d'Issarlès and the Lac de Saint-Front in the Velay and the Lac du Bouchet, near Cayres southwest of Le Puy.

The second phase of volcanic activity occurred during the late Pliocene-early Pleistocene and affected the Le Puy area, the upper valley of the River Loire and the Devès plateau southwest of Le Puy. Again it involved a combination of basaltic lava flows burying most of the plateau, together with intrusions, domes and cones. On later erosion some of the intrusions formed distinctive volcanic necks, the two most well known of which are within the city of Le Puy and are iconically crowned by chapel and statue (Fig. 8.4b). Already, here, the upper Loire had begun to incise its canyon into the underlying Hercynian granitic rocks. This second phase of volcanic activity partially filled these early canyons with lavas, which were later re-excavated to form the modern canyon.

There has been speculation as to whether this area, the Velay particularly, was glaciated during the late Pleistocene. I know of no specific evidence for glaciation, but there is evidence of intense periglacial activity, in the form of blockfields of randomly tilted angular blocks—perhaps remnants of a rock glacier. This would be evidence of intense periglacial processes sustained over a considerable time. It would be positive evidence for NO glaciation!

The Causses and the Southern Massif Central (Figs. 8.5 map, Fig. 8.6a–c): To the south of the Massif Central the Hercynian granitic basement is downfaulted and unconformably overlain by Jurassic limestones, forming one of the most important karst areas in France, the Grands Causses. This area is bounded to the east by the Cévennes (deeply dissected plateaux, primarily on Hercynian schists), to the west by the Aveyron plateau (on complex Hercynian structures), and to the south by the Luberon/Montagne Noire area (another zone of Hercynian rocks affected by later east-west Pyrenean structures).

The Causses plateaux have a whole range of karstic geomorphological features, from small-scale solutional features (Fig. 8.6a), on the Causse de Sauveterre. They also occur for example on the Larzac plateau (military land, on the D809 road southeast of Millau) to large scale underground drainage. Most spectacular of all is the Gorges du Tarn (Fig. 8.6b), a magnificent canyon dissecting and bisecting the area to the northeast of Millau. Very spectacular is the Pointe Sublime area, where the canyon is deep and narrow, and the channel is cut in bedrock. Further upstream though, there continues to be geomorphic interest. There are impressive incised meanders cut into the eastern edge of the Causses plateau near Quézac/Ispagnac. A little further upstream at Florac-Trois-Rivières it is possible to drive west out of the village up onto the lip of the plateau. From there, there are spectacular views up the valley of the Tarn towards its source at the foot of Mont Lozère (1699 m) within the Cévennes. South from Florac it is possible to access the Causses plateau (rock outcrops, solutional features etc.).

In the other direction from the Gorges du Tarn, the Tarn incision continues downstream of Millau (see Fig. 4.5c map), beyond the Millau viaduct, in the form of spectacular incised meanders.

To the southeast of Millau, near the eastern margin of the Causses plateau is another spectacular gorge, that of the River Vis. Unlike the Tarn, an Atlantic drainage, the Vis is a tributary of the Hérault, a Mediterranean drainage sourced within the Cévennes. The Vis gorges include the spectacular so-called "Cirque de Navacelles" (Fig. 8.6c) (see also Amber, Chap. 11 in Fort and André, Eds., 2014). This is not a cirque at all but a fascinating abandoned cutoff incised meander, involving subterranean karstic drainage. It is an isolated site, a long way from anywhere, but certainly worth the visit!

Gorges/Canyons/Incised Rivers within the Massif Central as a Whole   Throughout the preceding text on the Massif Central and especially within the descriptions of the highlighted sites, mention has frequently been made of the canyons, gorges and other incised river valleys. The depth of incision within the Massif Central is generally greater than that characteristic of the common post-Neogene incision elsewhere in France. The incisional morphology ranges from incised meanders, often with little or no floodplain development, to gorges and canyons. This is indicative of the sustained Neogene-Quaternary tectonic uplift of the Massif.

I thought it might help to list here the main canyons, gorges, and other incised river valleys within the Massif Central.

Loire Drainage
Upper Loire (from above Le Puy downstream virtually to Roanne)
Upper Allier (downstream to Brioude)
Sioule (downstream to Gannat)
Gironde drainage
Vezère and Corrèze (downstream to Brive-la-Gaillarde)
Dordogne (virtually the whole of the upper Dordogne downstream to Bretenoux)
Cère (especially between Aurillac and Bretenoux)
Célé (downstream to Figeac)
Truyère (downstream to the Lot confluence)
Lot (throughout, downstream to Capdenac)
Aveyron (from Rodez downstream to Laguépie—confluence with the Viaur)
Viaur (throughout, from Pont-de-Salars downstream to Laguépie: Aveyron confluence)
Tarn (from the upper Tarn above Quézac, through the Gorges du Tarn through Millau, downstream to beyond Ambialet)
Agout (downstream to Castres)
Mediterranean and Rhône drainage
Vis (throughout the reaches downstream to the Hérault confluence)
Hérault (downstream to Gignac)
Upper Gard/Gardon (both headstreams above Anduze)
Upper Cèze (downstream from the Cévennes to St-Ambroix)
Upper Chassezac and headwaters (downstream to Les Vans)
Upper Ardèche (downstream to Aubenas)
Doux (throughout)/Cance (downstream from Annonay to Rhône confluence)

Chapter 9
The Paris Basin

With the exception of the lowlands along the Rhône/Saône corridor and the Languedoc/Mediterranean coast the main lowland areas of France are the post-Hercynian sedimentary basins that lie between the various Hercynian massifs described in the previous chapters. These lowlands are: (1) the Paris basin, together with its extensions north into Picardy, west through Normandy, east towards the Vosges, and south to the Loire valley, and (2) the Aquitaine basin. I deal with the Paris Basin first (Chap. 9), followed by the Aquitaine basin (Chap. 10).

This Chapter is subdivided as follows (Fig. 9.1 map):

The Northern Chalk Plateaux of Picardy, Normandy, Perche.
The Eastern Scarplands.
The Central Paris Basin (The Île de France).
The Middle Loire Valley.

A. Harvey, *The Geomorphology of French Landscapes*, https://doi.org/10.1007/978-3-031-68490-6_9

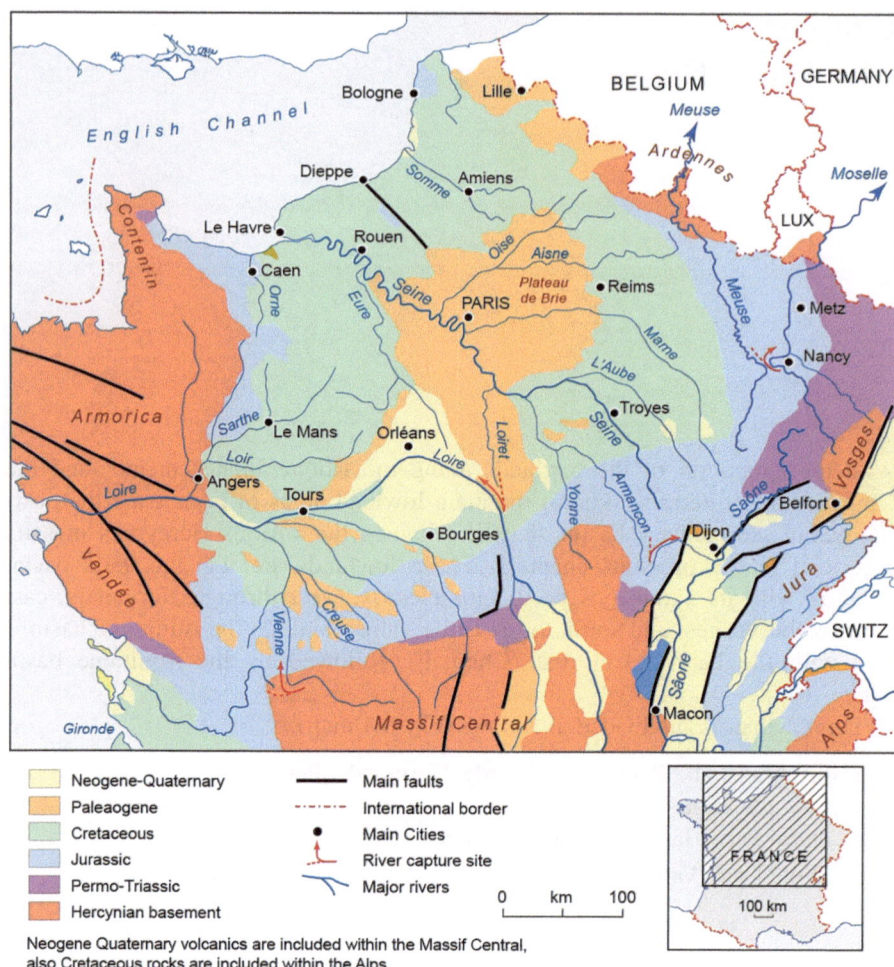

Fig. 9.1 Map of the Paris basin

The Northern Chalk Plateaux: Picardy, Normandy, Perche

The Geology of the Northern Chalk Plateaux

The geology (Fig. 9.2 map) of this area is dominated by the gently dipping, slightly flexured, Upper Cretaceous Chalk formation, locally capped unconformably by basal Palaeocene sands and clays. The very north of the area in the vicinity of Calais is formed of Quaternary marine silts. There are two areas of more pronounced structures exposing older rocks than the Chalk, both structures likely of "mid-Tertiary" (Alpine) age. The first is the Boulonnais anticline (north of Boulogne) exposing Lower Cretaceous rocks with a core of Upper Jurassic

Fig. 9.2 Map of the northern Chalklands

dominantly limestones. This anticline is essentially a small eastward cross-Channel extension of the Wealden anticline in England. The second area is the fault-aligned Bray anticline southeast of Dieppe, also exposing Lower Cretaceous and Upper Jurassic rocks below the Chalk. Again this structure is the cross-channel equivalent of a structure present in England, the anticline of the Isle of Wight. In the far west of the region near Caen is a low Chalk escarpment above the Jurassic rocks of western Normandy, which extend west to rest unconformably on the Hercynian rocks of the Cotentin peninsula (see Chap. 6). Locally, unconformably on an eroded surface across the Chalk, are patches of Palaeocene/Eocene sediments similar to those of the inner Paris Basin.

The Geomorphology of the Northern Chalk Plateaux: Inland

The relief of this area is dominated by extensive Chalk plateaux (Fig. 9.3a), whose elevation rarely exceeds 200 m. Only in eastern Normandy, in Picardy on the flanks of the Bray anticline and southeast of Le Tréport on either side of the Béthune valley, do elevations marginally exceed 200 m. This general elevation might relate to a former sub-Eocene surface exposed when the overlying Eocene sediments had been removed. Alternatively, the extensive plateau surfaces might relate to long periods of erosion during the Neogene, and could therefore be the equivalents of such surfaces on the Hercynian massifs. In exposures of the Chalk there are often deep red soils piped into the Chalk (see Fig. 3.3d). These would indicate formation over long periods of probably warm humid conditions. In southern England similar features have been attributed to sub-tropical weathering during the late Miocene and the Pliocene.

Fig. 9.3 The northern Chalklands. (**a**) The Chalk plateau near Vanault east of Laon. (**b**) The Seine at Les Andelys looking downstream. The incised meanders are trenched below the Chalk plateau (in this picture forming the skyline). Chalk cliffs form the valley wall on the outside of the incised meander (to the right)

There is one element of British Chalk landscapes that has been documented for the London basin by the classic work of Wooldridge and Linton (1955) that does not appear to have been recorded in relation to the Paris Basin. Wooldridge and Linton argue for the presence of latest Pliocene to basal Pleistocene marine sediments at elevations of c180 m on the Chalk margins of the London basin. I can find no authenticated mention in the French research literature of any equivalent sediments on the Chalklands of northern France. On the small-scale (1:1,000,000) geological map of France I can only find one locality where there are mapped sediments of Pliocene to Early Pleistocene age capping the Chalk. These are near Joigny on both sides of the Yonne valley gap through the Chalk escarpment at elevations of up to 270 m. I can find no mention of these in the literature, but I suspect they are more likely to be fluvial sediments, than marine sediments equivalent to those described in England. Maybe I have missed something. Maybe other such localities do exist but have not been described in the literature, or maybe the latest Pliocene-early Pleistocene marine conditions described by Wooldridge and Linton (1955) in parts of the London basin indeed did not extend into northern France. There is some evidence for this view, in that patches of the equivalent sediments have been mapped in southern Belgium, apparently as for those in the London basin, marking the approximate limit of marine conditions at that time.

Set below the plateau surfaces is the incised Quaternary drainage. With the exception of the Somme and the lower Seine, and to some extent the Orne in the far west of Normandy, the rivers and streams draining the Chalk plateaux are short streams, most in straight valleys oriented SE-NW towards the coast. There has been considerable research into the sequence of Pleistocene incision by these rivers from evidence preserved in the terrace sediments (see "Incision" section in Chap. 4, and references cited therein). The modern channels, where they have not been artificially straightened, are mostly small meandering channels, sharing the floodplains with managed wetlands. Of the two main exceptions, the middle Somme between Péronne and Amiens, is set in a broad open valley that contains incised valley meanders. Within these meanders the modern channel is anastomosing and includes small-scale channel meanders which are clearly underfit in relation to the valley bends (see Chap. 4). This type of channel is common elsewhere in France, but less common in these northern Chalklands. The Somme is sourced in Chalk springs near St Quentin before it heads north towards Péronne. Just south of St Quentin is the Oise, incised into the Chalk plateau. It is a Seine tributary, and heads southwest towards the central Paris basin (see later).

The other obvious main exception to the relative lack of incised meanders into the Chalk is the Seine itself below Mantes-la-Jolie. Here there are huge incised valley meanders (Fig. 9.2 map) within which the managed modern channel is mostly a single thread channel, broadly following the geometry of the valley bends. Locally, downstream of Jumièges, itself downstream of Rouen, there are hints of smaller dimension channel meanders. Elsewhere, especially around Les Andelys and upstream there are islands, remnants perhaps of a former anastomosing channel pattern. Also, in the Tancarville area there are remnants of now abandoned former large meander bends. Throughout this section of the Seine valley, preserved on the inside of the valley bends, there are terraces representing stages in the Pleistocene valley incision. To see some of the features described above one of the best places is Les Andelys (Fig. 9.3) upstream of Rouen, with the ruins of Château Gaillard preserved above the valley giving good views over the valley, a good reason for a touristic visit here anyway!

In the far west of Normandy, there is another exception to the general rule of few incised meanders, the River Orne, upstream of Caen. This area is west of the Chalk outcrop, occurring on Jurassic rocks that rest on the underlying Hercynian basement of the Armorican Cotentin peninsula (see Chap. 6). The course of the Orne switches between flowing on the basement rocks of Armorica and on the Jurassic cover rocks. Throughout the reach between somewhere near Argentan and near Thury-Harcourt it has incised valley meanders with the modern channel clearly misfit.

In most areas, including the edges of the Bray and the Boulonnais anticlines (Fig. 9.2 map) the outer margins of the Chalk plateaux are marked by distinct escarpments. The inner margins differ between the distinct scarps on the north and east of the inner Paris basin in Picardy and Champagne (see later this Chapter), and the much more subdued transitions elsewhere. The Chalk plateau continues south of the Seine valley into Perche. It is bounded there to the east by the Beauce plateau on the Miocene rocks of the inner Paris basin (see later this Chapter), and to the south by the middle Loire valley (see end section of this Chapter). On the western margin there is a weak escarpment above the Jurassic rocks that flank the Armorican massif in north-eastern Normandy. Perche is drained in the south by the Loir. The Loir and the Sarthe that drains the western margins of Perche, and the Mayenne, that drains eastern Brittany, are confluent with the Loire at Angers. The Sarthe and the Loir have incised valley meanders within which the modern meandering channels are markedly misfit. These are particularly well developed on the Loir downstream of Châteaudun and also downstream of Vendôme.

The Northern Coasts

Perhaps the most interesting and varied geomorphology of this region is its variety of coastal landforms. East of Calais the coast is dull, low dunes backed by reclaimed saltmarsh, then industrialised areas east to the Belgian border. A few km west of Calais is the Boulonnais anticline, culminating in cliffs in Jurassic limestones at Cap Gris-Nez. South of Boulogne it is a mixed coast, mostly sandy beaches backed by dunes, with a local interruption by a spit and the Canche estuary at Le Touquet. South of Berck-Plage is the muddy estuary of the Authie with its saltmarshes. Then there is a continuation of the beach and dune coast to the large estuary of the Somme between Le Crotoy and St Valery-sur-Somme.

Beyond the Somme estuary, reclaimed saltmarsh gives way to low Chalk cliffs from Ault through Le Tréport to Dieppe. For most of this stretch the low cliffs are footed by a sandy beach, but near Dieppe the sand is replaced by a wave-cut platform in the Chalk. Beyond Dieppe cliff height increases to around 50 m, but the most spectacular cliffed coast lies further west, beyond Fécamp and particularly around Étretat (see Fig. 1.3c). For most of this stretch of coast cliff elevation exceeds 100 m. The cliffs are vertical. Most spectacular of all is the area around Étretat itself, which includes the (Monet-painted, and the often photographed) natural arches of the Falaise d'Aval. Relatively high Chalk cliffs continue to Cap de la Hève near Le Havre on the corner of the Seine estuary.

Beyond the Seine estuary, beyond Trouville are low cliffs in Upper Jurassic rocks, capped inland by Cretaceous rocks. Further along, towards Dives/Cabourg, are degraded cliffs in Jurassic rocks. Beyond the Orne estuary low degraded cliffs in Jurassic rocks continue, in many places fronted by a sand beach. One of these is Omaha beach (one of the 1944 Second World War Allied landing beaches). At the end of this stretch, before the Marais Cotentin, the degraded cliff height increases to up to 50 m.

The Eastern Scarplands: Champagne, Lorraine, Northern Burgundy

The Geology of the Eastern Scarplands
In this area the sequence of Mesozoic rocks forms a series of arcuate belts aligned more or less north-south. From east to west these belts are: (i) Triassic sandstones flanking the Vosges (see Chap. 7); (ii) The dominantly limestone Jurassic sequence of western Lorraine, including the Côtes de Moselle and de Meuse; (iii) the Cretaceous sequence in Champagne including the Lower Cretaceous, Argonne ridge and the Upper Cretaceous Chalk plateau; (iv) the Palaeocene-Eocene escarpment of the Falaise de l'Île de France (Figs. 9.4 map, 9.5).

Geomorphology of the Eastern Scarplands
Figure 9.5 illustrates how the gross topography of successive escarpments and intervening vales/lowlands reflects the underlying geological sequence. The east-west sequence of east-facing escarpments culminates in western Champagne in the escarpment of Eocene Calcaire de Brie limestone (the Faliase de l'Îsle de France).

The Calcaire de Brie escarpment can be traced from Laon, the city spectacularly sited on top of the scarp, south to Reims in Champagne. It is on the lower slopes of this escarpment, to the south of Reims, around the Montagne de Reims and where the escarpment is cut by the Marne valley near Épernay, that some of the most famous Champagne wine is produced. Extending east and southeast from the base of this escarpment is the Chalk plateau (Fig. 9.3a). This plateau is about 30 km wide and rises gradually in elevation from less than 100 m near Chalons-en-Champagne

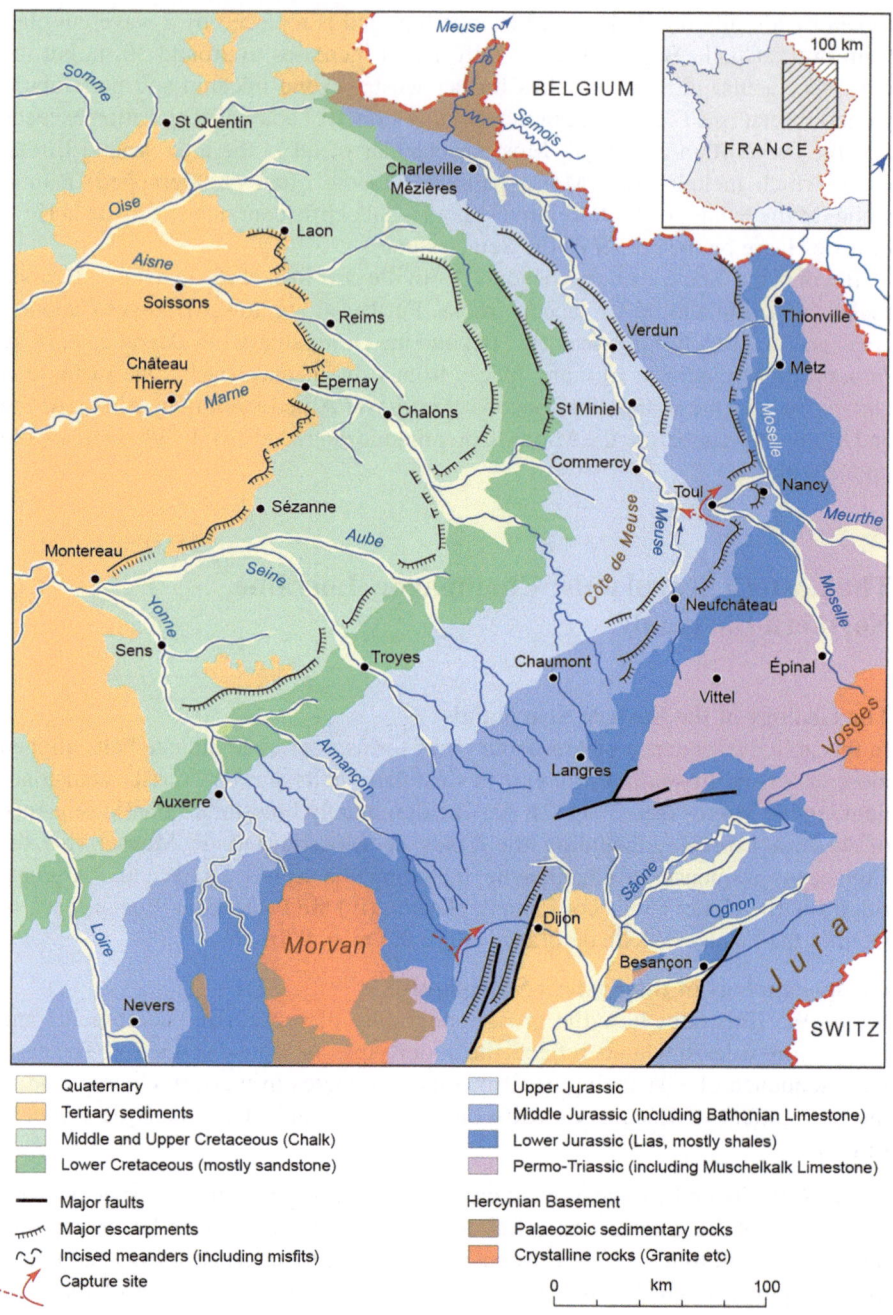

Fig. 9.4 Map of the eastern scarplands

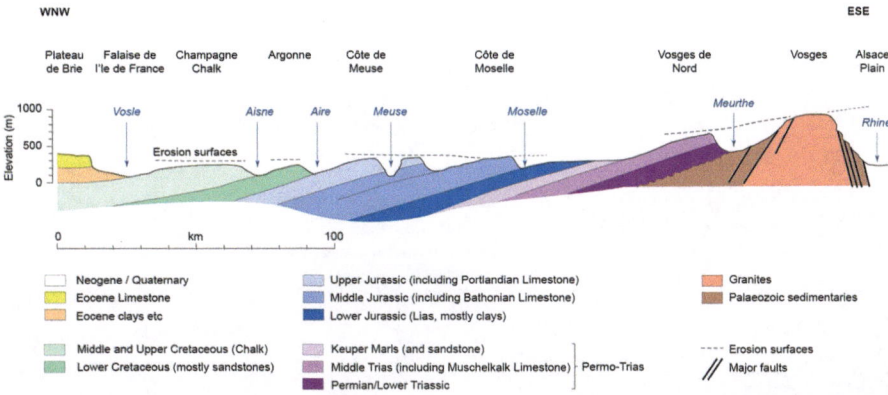

Fig. 9.5 Geological and generalised topographic section across the eastern scarplands

to about 250 m on the crest the Chalk escarpment. The Chalk plateau arcs south-westwards from near Troyes towards Joigny, forming an escarpment with elevations of up to about 270 m. Near Joigny on either side of the Yonne river gap the Chalk is capped by Plio-Pleistocene sediments of uncertain origin, described in the preceding section of this Chapter (northern Chalklands). From the Yonne gap westwards towards Montargis the plateau loses some elevation and becomes less distinct a feature as it is increasingly overlain by Eocene sands and clays. Throughout the length of its outcrop the escarpment is broken by river gaps (all of which are parts of the Seine system), from north to south: the Aisne (north of Reims); the Marne (near Vitry-le-François); the Aube (north of Troyes); the Seine itself (at Troyes); and the Armançon/Yonne (near Joigny).

Beyond the Chalk outcrop and beyond the base of the Chalk escarpment is a clay vale, followed by another east- and southeast-facing escarpment, that of the Lower Cretaceous sandstones, forming the wooded Argonne ridge (the equivalent formation to the English Lower Greensand). Further east is a relatively narrow lowland vale, with a floor at about 150 m, beyond which is a complex area of multiple scarps in Upper Jurassic rocks, culminating in the Côtes de Meuse (Fig. 9.6b). This is formed of an Upper Jurassic limestone, perhaps the equivalent of the English Portland stone.

Interestingly the Côtes de Meuse scarp lies to the east of the Meuse valley itself. Elevations of the scarp crest reach almost 400 m whereas the Meuse valley near Verdun is at elevations of less than 200 m. The belt of Upper Jurassic rocks can be traced southwestwards through Tonnerre on the River Armançon towards the upper part of the Middle Loire valley (see the final section of this Chapter). However the scarp forms become much less distinct westwards from the Armançon valley. In Lorraine, Eastwards from the Côtes de Meuse, after a narrow vale, there is the dip slope of the middle Jurassic (Oolite) escarpment of the Côtes de Moselle. This scarp rises above the Moselle valley to elevations of almost 400 m. It can be traced south into the Langres plateau, the gateway to the Saône basin, and then westwards into

Fig. 9.6 The eastern scarplands. (**a**) The 'Falaise de l'Île de France': The escarpment of the Palaeocene-Eocene Calcaire de Brie marks the boundary between the Inner Paris basin on Cenozoic rocks, and to the east the plateau on Cretaceous Chalk. Here, just north of Épernay, the southeast-facing scarp slopes provide the 'terroir' for some of the most exclusive Champagne wines. (**b**) The Côtes de Meuse escarpment: Upper Jurassic limestone near Verdun, Meuse

Burgundy where it forms the Côte d'Or escarpment (see Chap. 12). In Lorraine it can be traced south, eventually giving way to Triassic rocks on the flanks of the Vosges.

Bevelling the plateaux and escarpments are Neogene (?) erosion surfaces, in some areas carrying relict *terra rossa* soils (Fig. 3.3d). The dissection below them relates to the Pleistocene (ie. to the last 2 million years or so). There was no Pleistocene glacial ice in this region but the period was marked by alternations between periglacial and more temperate periods. The periglacial periods were characterised by permafrost and by high runoff.

Some of the incised valleys show incised meanders, within which the modern rivers are "misfits", though this phenomenon is not as well developed as it is in some other areas of France. Incised meanders are not well developed on the Oise except in a few localities, nor on the Aisne, the Marne, nor the Moselle, but there is moderate development on the meandering channel of the Meuse. Further south the valleys of the upper Seine and the Aube are more or less straight, but the modern channels have tortuous meanders. Within the Yonne system both the upper Yonne and the Armançon have incised meandering valleys, within which the modern streams are misfit.

In the east of the area is one of the classic cases of river capture, the Pleistocene capture of the former upper Meuse by the Moselle at Toul (Fig. 9.7 map), within valleys cut into the Côtes de Moselle. The Palaeo-Moselle upstream of Nancy was

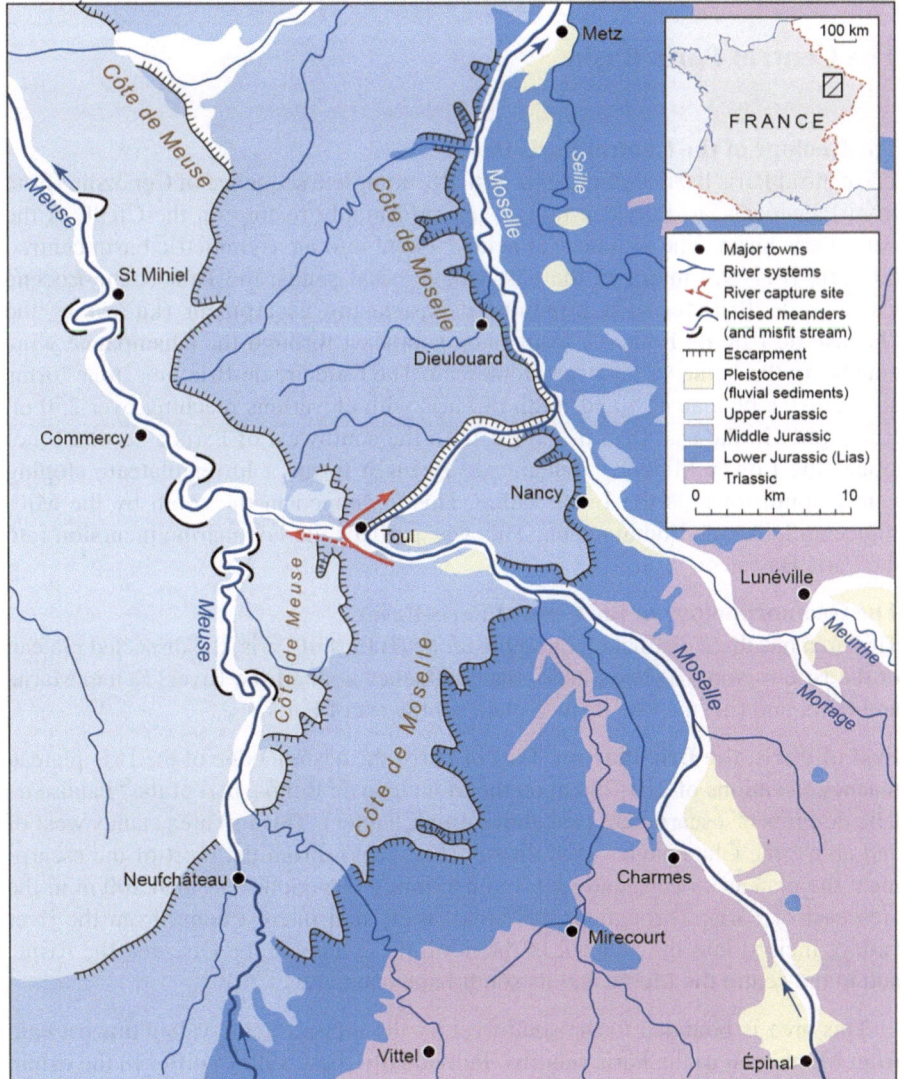

Fig. 9.7 Map of the Meuse/Moselle capture site at Toul, Lorraine

a tributary of the Meuse. In mid-late Pleistocene it was captured by the lower segment of the modern Moselle, presumably as a result of the lower base level provided by that river.

To the southeast of the area, near Bligny-sur-Ourche in Burgundy, there is evidence of another stream capture, that of the Seine/Armançon system headwaters by the aggressive drainage of the Ourche (see Chap. 12). The Ourche is a tributary of the Saône system. This capture presumably occurred in response to the tectonically induced steep gradient of the Ourche on the margins of the Saône graben (see Chap. 12).

The Central Paris Basin

The Geology of the Central Paris Basin
The central Paris basin comprises a virtually complete sequence of Cenozoic strata from Palaeocene to Pliocene in age, unconformably resting on the Chalk of the outer Paris basin. The sequence is gently folded into an asymmetric basin centred on Paris (Fig. 9.8 map). In the east, above basal sands, the Palaeocene-Eocene Calcaire de Brie forms a pronounced east-facing escarpment (known as the "Falaise de l'Île de France") from Laon southeast through the Champagne wine country to the Seine valley near Montereau. The Calcaire de Brie limestone forms an extensive plateau east and south of Paris with elevations reaching over 250 m. It is higher in the east than in the west. To the southwest of Paris a second limestone (the Lower Miocene Calcaire de Beauce) forms a lower plateau, sloping gently south towards the Loire valley. This is capped in the south by the Mio-Pliocene Sables de Fontainebleu. This was perhaps the last marine incursion into the Paris Basin.

The Geomorphology of the Central Paris Basin
This area has three distinct topographic units: (i) East of Paris, the dissected plateau of the Brie region; (ii) Paris itself, the confluence zone of the Rivers Seine, Marne and Oise; and (iii) the low Beauce plain southwest of Paris.

East of Paris, the Brie Plateau East of Paris, the eastern edge of the Brie plateau reaches elevations of 250–280 m on the Montagne de Reims, part of the "Falaise de l'Île de France" escarpment (see above, this Chapter). This hill area stands west of and above the Champagne wine district (Fig. 9.6a). From the crest of the escarpment the plateau slopes gently west to maximum elevations less than 200 m in the area east of Paris. Throughout this area variety and interest come from the river valleys incised into the plateau. In the north these include the Oise and the Aisne, and in the centre the Marne and its south-bank tributaries.

This area is bounded to the southwest by the incised Seine valley downstream from Montereau to the Paris suburbs. Individually these valleys differ in the extent

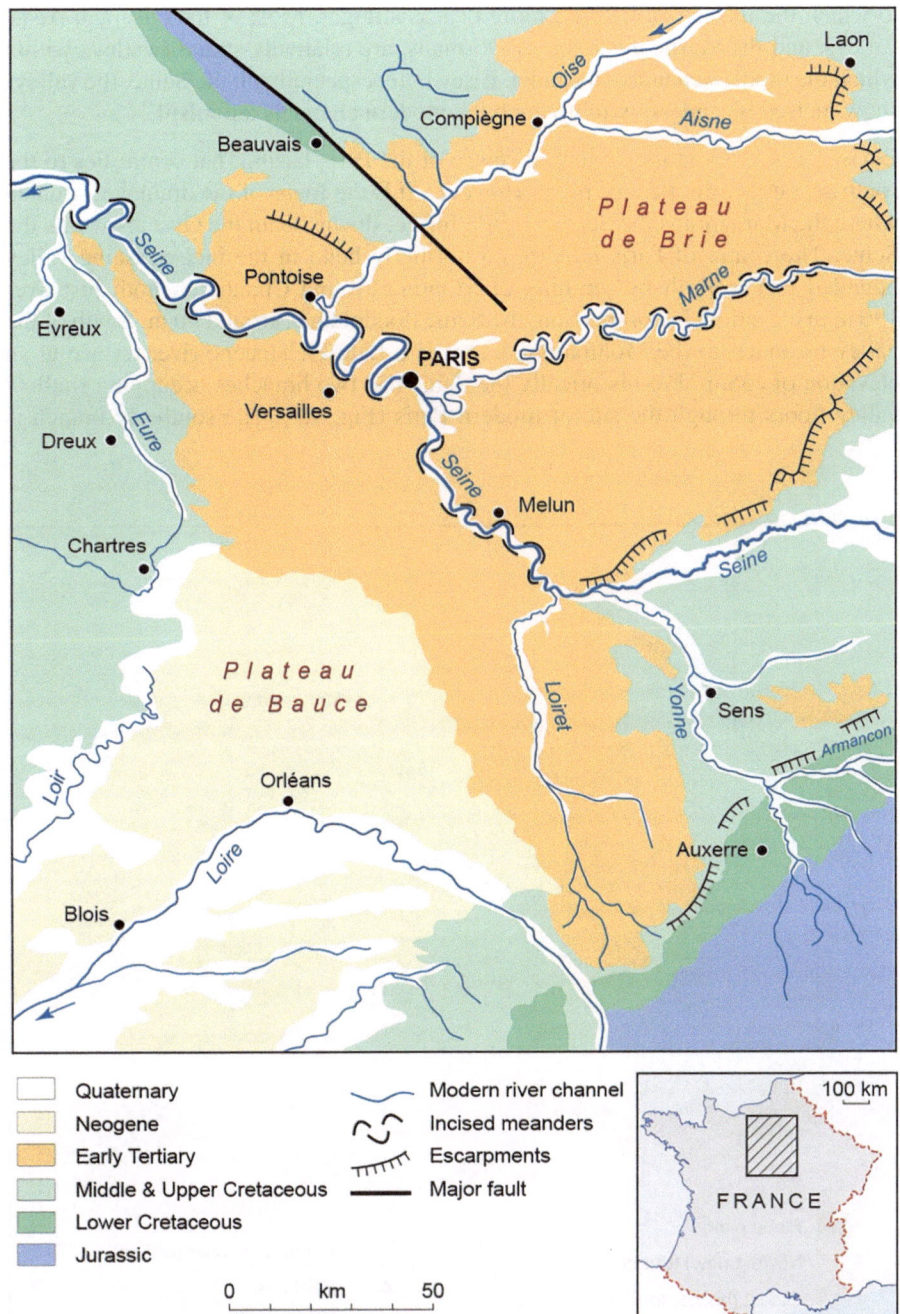

Fig. 9.8 Map of the central Paris basin

Quaternary

Neogene

Early Tertiary

Middle & Upper Cretaceous

Lower Cretaceous

Jurassic

Modern river channel

Incised meanders

Escarpments

Major fault

FRANCE

100 km

to which the incised valleys are more or less straight. Most of the Oise and Aisne valleys, and the Marne upstream of Dormans, are relatively straight valleys within which the modern channels meander. Elsewhere, especially on the Seine, the valleys show incised meanders, within which the modern channels are misfit.

Paris   Paris is not at the structural centre of the Paris basin. That centre lies to the southwest under the Beauce plain. However, it is the focus of the drainage, situated where the Marne joins the Seine. A little further downstream the Oise also joins the Seine. The centre of Paris is within a northerly bend in the incised Seine valley bounded to the north by the hills of Montmartre and Chaumont, both just over 130 m in elevation. In comparison, the Seine floodplain is around 30 m. South of the valley meander, in the Montparnasse area is a Late Pleistocene river terrace at an elevation of c55 m. Pre-historically the Seine had two branches occupying shallow valley floors through the site of modern Paris (Fig. 9.9). The southerly branch is

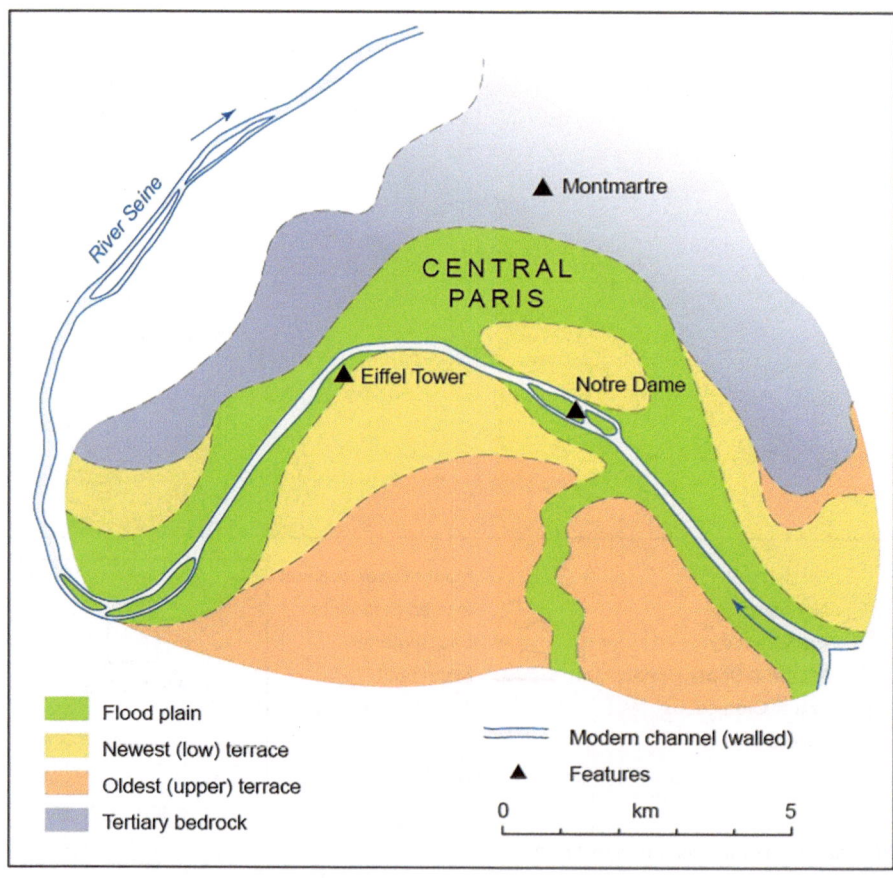

Fig. 9.9 Map of the geomorphic features of Paris

now followed by the main river, and the northerly branch runs through the sites of the Bastille and Saint-Lazare. The two branches rejoined in the Champs de Mars area. There was a low island between these two branches, the site of the modern city centre. With the growth of the city, the southern course became the main course, with the channel later to be confined within the modern channel walls (finally established during the eighteenth Century).

This situation has led to the modern city having a flood problem (see Fort et al., Chap. 2 in Fort and André, Eds. 2014). Situated downstream of the Seine/Marne confluence, there is potential for a flood hazard generated from a wide area of France. This, together with the confinement of the channel by the floodwalls, accentuates the likelihood of flooding. The highest floods on record occurred in 1910, when much of the northern palaeochannel area was flooded (including Saint-Lazare station). Another major inundation occurred in 1924, and in recent years there have been significant floods in 2016 (generated as a flash flood from a summer storm), and more recently in January 2018 (generated from widespread and long duration heavy winter rainfall).

Beauce, SW of Paris The Beauce plain extends southwest from the inner Paris basin to the middle Loire valley between Orléans and Blois. It is of very modest relief rarely exceeding 160 m in elevation. It is bound to the south by the middle Loire valley (see below, next section of this Chapter), to the west by the upper Loir (see also below, next section of this Chapter), and to the east by the Loing, tributary to the Seine at Fontainebleu.

The Loing valley marks a structural feature, basically defining the eastern edge of the Beauce plain. The Chalk to the east of the flexure dips westwards in a monoclinal structure, perhaps modified by faults. Until some stage during the late Pliocene or early Pleistocene(?) the Loing provided the outlet for the Loire. So at that time the Loire was part of the Seine system. This is evidenced by ancient river gravels within the Loing valley containing sediments of Massif Central origin that can only have been derived from the palaeo-Loire (see Chap. 3, see also Tourenq and Pomerol 1995). The modern Loing, a small river, flows north through a broad ill-defined valley, that was the early Pleistocene valley of the Loire.

The Middle Loire Valley

The Geology of the Middle Loire Valley
Between the northern edge of the Massif Central and the middle Loire valley are successive belts of Jurassic then Cretaceous rocks (Fig. 9.10 map). The Jurassic rocks extend southwest through the Poitou gate into the north of the Aquitaine basin (see Chap. 10). Unconformably overlying these Mesozoic rocks are patches of Palaeocene/Eocene rocks, similar to those elsewhere in the Paris Basin.

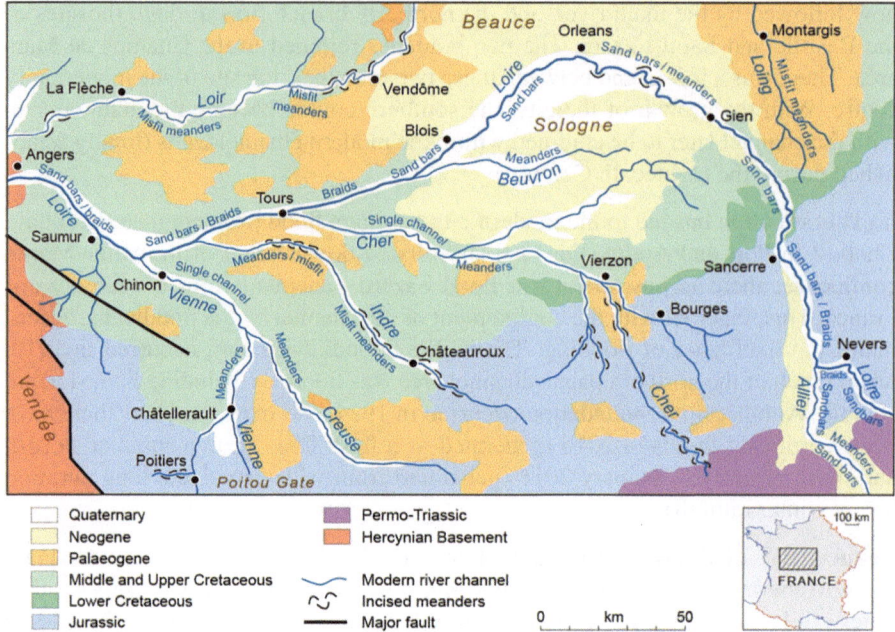

Fig. 9.10   Map of the Middle Loire valley (including the Loire-Loing capture zone)

South of the modern River Loire within the great bend of the modern river is a suite of Upper Miocene, Pliocene and early Quaternary dominantly fluvial sediments. These form an ill-drained almost fan-like feature with a radial drainage pattern (the marshes of the Sologne, Fig. 9.10 map). This feature relates to the initial post-palaeo Loire-Loing drainage. The modern Loire developed northeast of this, flowing towards Orléans.

The Geomorphology of the Middle Loire Valley

Away from the margins of the Massif Central the relief of this area is very subdued. Even the Jurassic and Cretaceous escarpments are only weakly developed. The distinctive relief features all relate to the river valleys and the modern river channels. Upstream of their confluence near Nevers the two main headstreams of the Loire, the upper Loire itself and the Allier, drain the eastern part of the Massif Central in Auvergne via downfaulted rift basins, the Roanne/Loire and Limagne/Allier basins respectively (see Chap. 8).

From Roanne downstream to Decize the upper Loire has an active meandering channel in a fairly wide floodplain, within which there are occasional abandoned meanders. Then from Decize downstream to the Allier confluence near Nevers, although still a meandering channel, there are more active gravel bars and modern meander scrolls. Interestingly, the extent of modern channel migration in the

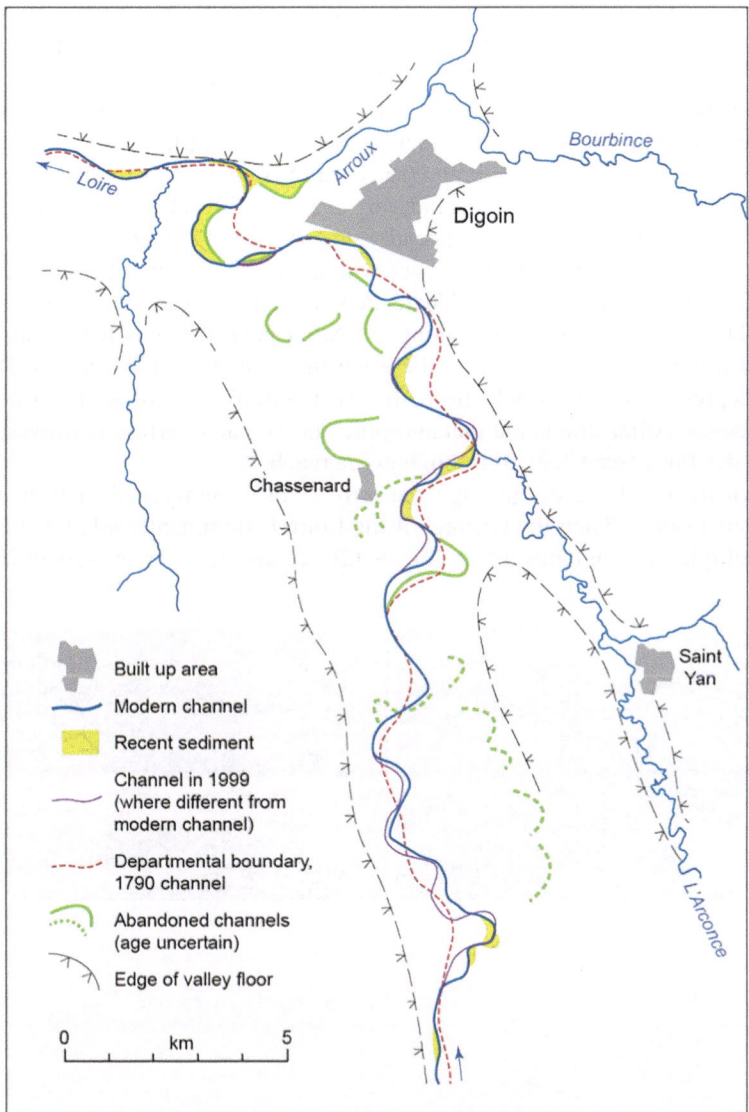

Fig. 9.11 Map of the River Loire south of Digoin: meander migration

meandering reach between Marcigny and Digoin (Fig. 9.11 map) can be assessed by
the discrepancy between the Departmental border (presumably located down the
channel at the time it was fixed in the eighteenth century?), and the modern channel
position. On most bends there appears to have been a lateral and downstream shift
of about 500 m.

The main tributaries from the east draining western Burgundy, the Arroux and the Bourbince, also have meandering channels within their floodplains. Downstream from Digoin the Loire has a broad, dominantly meandering channel, but further downstream the river begins to braid (Fig. 9.12a) before it becomes fully braided near Nevers in the Allier/Loire confluence zone (Fig. 9.12b).

The Allier from Clermont-Ferrand (see Chap. 8) downstream to the Loire confluence near Nevers, has a mostly meandering channel with active point bars showing evidence of channel migration, especially downstream of Varennes-sur-Allier. For the last 10 km or so before its confluence with the Loire the Allier has a more or less a straight channel with braid bars. There are two main tributaries to the Allier in this reach. The Dore, from the east joining the Allier near Thiers is fed by an incised channel through the granite terrain between the Limagne/Allier and the Roanne/Loire depressions. The Sioule from the west joining the Allier downstream of Varennes-sur-Allier drains the metamorphic and volcanic terrain northwest of the Chaîne des Puys (see Chap. 8) through gorge reaches.

Downstream of Nevers through Sancerre (wine country!), then from Cosne-Cours-sur-Loire to Gien the channel of the Loire is dominantly a braided channel with multiple sand and gravel bars (Fig. 9.12b see also Gautier and Grivel 2006). It

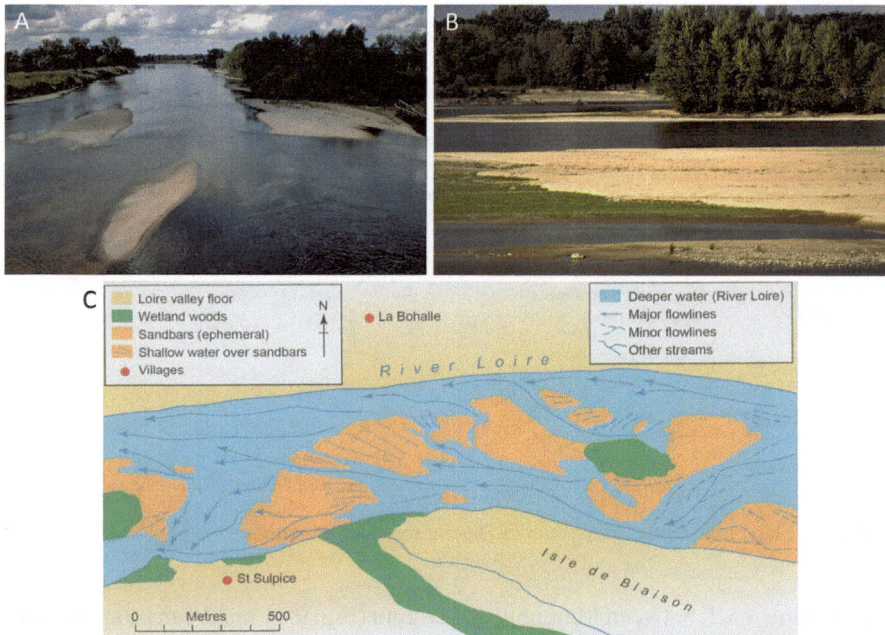

Fig. 9.12 The Middle Loire valley: Digoin—Nevers—Orléans—Angers (**a**) The Loire at Gannay, downstream of Digoin: Within this upper reach there is incipient braiding, manifested by mid-channel sand bars. (**b**) Fully braided channel of the Loire at the Loire/Allier confluence downstream of Nevers. (**c**) Map of the channel of the Loire upstream of Angers. The alternating sandbars are indicative of incipient braiding

is from this area that prior to the early Pleistocene the Loire headed north into the valley of what in now the Loing and ultimately into the Seine drainage. Capture took place in the vicinity of Briare (Fig. 9.10 map; see also Fig. 3.4 map) to the west into the Sologne area, followed by successive migrations northwards to form the great Orléans bend in the modern Loire valley.

Beyond Gien as far as Orléans the modern channel is a wide meandering channel with point bars within a fairly wide floodplain. However, the channel is not far from the braiding threshold, as evidenced by numerous sand/gravel bars and occasional islands. From Orléans through to Blois (into Loire "Château" country) the Loire has a fairly straight channel with occasional sand and gravel bars, but from there through Amboise to Tours it is a true braided channel with multiple bars and indeed some vegetated islands. This style of channel continues through Saumur to the Angers area (Fig. 9.12c map), where the lower Loire enters the Armorican realm transverse to the rocks and structures of the Armorican, Vendée massif (see Chap. 6).

Upstream of Nevers there are tributaries joining the Loire from both east and west. They are fed by the northernmost parts of the Massif Central, and have dynamic, active meandering channels (eg. the Bersbre, near Dompierre Fig. 9.13a). Beyond the Allier confluence near Nevers there are no significant tributaries to the Loire until way beyond Orléans. From there to Amboise all the main tributaries come from the south. The first of these tributaries of any size is the Beuvron, a small stream joining the Loire southwest of Blois and draining the swampy Sologne. Then in sequence the main tributaries are: the Cher (Fig. 9.13b) flowing through the south of the city of Tours, but joining the Loire some way downstream; the Indre, joining the Loire near Langeais; the Vienne, joining the Loire upstream of Saumur with its eastern tributary, the Creuse; then finally the Thouet a small stream joining the Loire at Saumur itself.

These rivers are all relatively low-gradient, relatively stable meandering rivers (Fig. 9.13b) (Vayssière et al. 2020). The Cher has a meandering pattern in a relatively wide floodplain throughout most of its length downstream from its incised headwaters in the northern Massif Central upstream of Montluçon. Otherwise there is little evidence of the development of incised valley meanders that might be expected of rivers in central France. A local exception is a small south bank tributary to the Cher, the Fouzon south of Selles.

The Indre is a much smaller river than the Cher. It rises just to the north of the Massif Central (in northeast Limousin). For most of its course it is a diminutive misfit meandering stream within a fairly wide (non-meandering, nor deeply incised) valley floor. Only locally are there exceptions. Near Châteauroux in its middle reaches there are relatively poorly incised valley meanders within which the modern Indre is misfit. Also in its downstream reaches near Montbazon, south of Tours, there is a hint of incised valley meanders.

The Vienne, together with its east bank tributary the Creuse, is a much larger river system. Both the Creuse and the Vienne head in Limousin in the northwest of the Massif Central. Both have incised meandering valleys in that area (see Chap. 8). Within that area too, the Vienne appears to have augmented its upper drainage area by the capture of the former Charente headwaters (see Chap. 8). The Creuse in its

Fig. 9.13 Loire basin: Southern tributaries. (**a**) River Besbre, a left bank tributary of the Upper Middle Loire, west of Digoin: Meandering channel near Dompierre sur Besbre. (**b**) The River Cher, near Villefranche-sur-Cher: A single-thread low-gradient meandering channel on the southern margins of the Sologne wetlands

upper reaches, within the Massif Central, flows in irregular incised valley bends. Once beyond the margins of the Massif Central it flows in an irregularly sinuous channel, locally meandering, within a relatively straight floodplain.

The Thouet rises on the eastern flanks of the Vendée granitic massif (see Chap. 6). On the margins of the massif it has an incised meandering valley, within which the modern meandering channel is misfit.

Highlights: The Paris Basin

The Northern Chalk Plateaux (Pas de Calais, Picardy, Normandy, Perche) *Cap Gris-Nez* Cliffs in Upper Jurassic limestones. The Boulonnais anticline, the continental extension of the English Wealden anticline, brings up Upper Jurassic rocks (dominantly limestones) within the otherwise monotonous Chalk-dominated landscapes of the Pas de Calais. The Jurassic limestones are exposed as coastal cliffs at Cap Gris-Nez.

Somme Estuary at Le Crotoy Coastal depositional landscapes, including extensive saltmarshes, occur here. This lowland coast from Berck-Plage south to Cayeux-sur-Mer includes a range of coastal depositional morphology. This range includes sand beaches and associated sand spits at Berck-Plage, and coastal dunes at Fort-Mahon and Saint-Quentin-en-Tourmont. The two estuaries, the Authie in the north and the much larger Somme in the south, are characterised by mudflats and saltmarshes, used for the cultivation of shellfish.

Somme Valley Between Péronne and Corbie Incised meanders plus a Pleistocene incisional sequence are preserved within the river terraces. (A visit to this area could be combined with a visit to the First World War Battle of the Somme war graves nearby). There are well developed incised valley meanders on the River Somme cut into the Picardy Chalk plateau west of Péronne. Evidence for the incisional sequence, Pleistocene in age, is preserved in the river terrace sediments (see Antoine et al. 2007) occurring mostly on the insides of the bends. The Holocene valley floor would have had an anastomosing channel within well vegetated wetlands. The modern valley floor and channel which retain some of these characteristics is fairly heavily managed. The features of the valley can be seen between Vaux and Bray. There is viewpoint over the valley at (Belvédère de) Vaux.

Étretat Spectacular Chalk cliffed coast (see also Costa, Chap. 4 in Fort and André, Eds. 2014). The north Normandy coast from near Le Tréport all the way westwards almost to Le Havre, is a cliffed coast cut in Chalk. At the eastern end of this area the cliffs are generally about 50 m high, but in the west average about 100 m. Much of this coast is actively eroding. In places the cliffs are fronted by a bare wave-cut platform, otherwise by shingle beaches of flints. There are many localities that would justify a visit, but perhaps the best known and most spectacular is Étretat (see Fig. 1.3c). Here the erosional forms are spectacular, and include stacks and a natural arch—scenery that has inspired artists, particularly Gustave Courbet and Claude Monet (see Chap. 1).

Les Andelys (Château Gaillard) incised meanders of the Seine, cut in Chalk (Fig. 9.3b) (See also Peulvast et al., Chap. 3 in Fort and André, Eds. 2014). Downstream from Mantes-La-Jolie towards Rouen, the Seine valley is incised into the Chalk plateau. The plateau itself is at around 150 m in elevation, the river channel at about 15 m. The valley form is one of spectacular incised meanders, the outsides of the bends cutting into the Chalk, and the insides of the bends preserving terrace sediments that record the Pleistocene incisional history of the valley. The modern natural channel would be a multiple channel, with numerous linear islands. However, it is managed for navigation and therefore is at best only partly natural. One of the best locations to see the elements of this landscape is Château Gaillard at Les Andelys (combine a geomorphic visit with a cultural/historical visit). An additional site well worth a visit is the valley downstream of Rouen, between Tancarville and Pont-Audemer, where the Marais Vernier occupies a Holocene abandoned meander.

The 1944 Normandy Landing Beaches (Juno, Gold, Omaha, Utah), North of Bayeux Low cliffs in Middle and Upper Jurassic rocks. This is the area of the 1944 allied Normandy landings, and a geomorphological visit to the area could be combined with visits to these historical sites. In the east between Deauville and Dives the coast is one of low cliffs in Upper Jurassic rocks, mostly marls. The cliffs are fronted by sandy beaches, forming a spit at Dives. At Villers the cliffs are in gullied Oxfordian clays. From the Orne estuary at Ouistreham (more sand spits) to Arromanches (Juno and Gold beaches), there are initially sandy beaches, then beaches backed by low cliffs in Upper Jurassic limestones. This terrain continues west to Grandcamp-Maisy (through Omaha beach), then beyond the Carentan embayment to the coast on the eastern flank of the Cotentin peninsular (Utah beach, see Chap. 6) which comprises sandy beaches fronting low stable dunes.

The Eastern Scarplands—Escarpments Champagne Vineyard Slopes at Verzy On the scarp of the "Falaise de l'Île de France" south of Reims two important Champagne wine villages, Verzy and Verzenay, sit on the edge of the Montagne de Reims below the escarpment of the Palaeocene-Eocene Calcaire de Brie. These rocks are unconformable on the underlying Cretaceous Chalk. They form the forested plateau behind the villages. On the slopes below the scarp edge the Chalk is mantled by Pleistocene periglacial sediments, which form the basis for the vine growing. On the road between the two villages are excellent viewpoints over the vineyard-covered slopes of the escarpment to the valley floor of the Vesle (beyond the TGV line and the motorway). Beyond are the beginnings of the Chalk plateau of the dominantly wheat growing "Champagne Pouilleuse" (dusty!).

Escarpments in Cretaceous and Jurassic Rocks in Champagne, Lorraine and Northern Burgundy Going down the geological sequence from Upper Cretaceous to Lower Jurassic the major rock groups form successive (west to east) arcuate belts of country (Fig. 9.5). Beyond the Palaeocene-Eocene Calcaire de Brie, which forms the "Falaise de l'Île de France" scarp (see Verzy above), is the Upper Cretaceous Chalk. The Chalk forms the low plateau of the "Champagne Pouilleuse" (Fig. 9.3a), bounded on the east and south by the (not very pronounced) Chalk escarpment, above the Lower Cretaceous sands and clays of the "Champagne Humide" (damp!).

The Chalk escarpment can be traced from the southwest of Troyes (between Troyes and St Florentin) to the east of Troyes (between Troyes and Vitry-le-François), then northeast through Ste. Menehould to Semuy (on the Aisne). Where the Chalk is capped by Cenozoic rocks, the scarp top tends to be wooded (as it is to the southwest of Troyes) otherwise it tends to be wholly in cropland.

On the whole the "Champagne Humide", developed on Lower Cretaceous sands and clays, tends to be a lowland vale. However a locally more resistant sandstone (the equivalent of the English Lower Greensand) forms the prominent scarp of the Argonne to the northeast of Ste. Menehould.

Further east the Upper Jurassic limestones form the scarp of the Côtes de Meuse (Fig. 9.6b), which has an interesting relationship with the River Meuse itself. For most of its course the river valley lies behind the scarp (ie. to its west), but in the most upstream reaches above Neufchâteau, and in the most downstream reaches from near Stenay, the river valley is to the east of the scarp. This lack of accordance between structure and drainage pattern has nothing to do with the capture site at Toul/Nancy (see below), and suggests superimposition or antecedence of the original drainage during the Neogene? (but I do not know the details). The scarp itself is well developed west of Mouzon where it lies to the south and west of the river. After the gap at Mouzon and south of Mouzon it lies to the east of the river.

Further east still, and going down the geological sequence, is another major scarp, that of the Middle Jurassic limestone, the Côtes de Moselle. It forms a very distinct escarpment which can be traced northwards from the Châtenois area east of Neufchâteau, passing immediately west of the city of Nancy. Southwest of Nancy the Moselle cuts through the escarpment first heading west towards Toul, the capture site (see below), then turns east to Pompey, north of Nancy (Fig. 9.7 map). From there north the scarp is a pronounced feature to the west of the river through Pont-à-Mousson. The river then passes west of Metz, and then through Thionville, and from there it runs due north to the Luxembourg border. South of Thionville access and viewpoints are not as easy as they could be, partly because of the urban and industrial development of the Moselle valley itself, but the scarp is accessible north of Thionville, near Guentrange, where it forms a double escarpment.

The Middle Jurassic limestone scarp forms the Luxembourg border, then the Belgian border from near Longwy to north of Montmédy. It is breached by the Meuse north of Mouzon and from there west it more or less merges with the Côtes de Meuse to the south of Sedan and Charleville-Mézières. To the east of the middle Moselle valley, the terrain is initially on Lower Jurassic (Liassic) then Triassic rocks on the flanks of the Vosges (see Chap. 7).

Highlights—The Eastern Scarplands—River Valleys *Meuse/Moselle Capture Site at Toul* (Fig. 9.7 Map) Previously (at some stage during the Pleistocene), the upper Moselle fed through a gap through the Côtes de Moselle at Neuves-Maisons southwest of Nancy (as it does now). It then flowed through a gap in the Côtes de Meuse at Toul to flow into what is now the middle Meuse valley in the vicinity of Commercy. The present Meurthe, upstream of Nancy, was the main headstream of the middle Moselle. A left bank tributary of the (then) Meurthe below Nancy cut back from Pompey through the Côtes de Moselle to intercept the proto Moselle/ Meuse at Toul, thus capturing the upper Moselle and diverting it into the Middle Moselle. In this way it beheaded the Meuse and left an abandoned gap through the Côtes de Moselle at Toul (Fig. 9.7 map). The field evidence for this sequence is clearly demonstrated by the landforms in the Neuves-Maisons, Toul, and Pompey areas including the palaeo-Meuse valley prior to capture by the Moselle and the capture site at the Toul gap. The modern Meuse between St Germain-sur-Meuse and Commercy has incised valley meanders cut into upper Jurassic limestones, within which are tortuous modern misfit meanders. (Take the D400 and D36 backroads between Toul and St Germain-sur-Meuse.)

Champagne Vineyard Slopes Above the Marne Valley at Épernay, a Panorama   At Épernay the River Marne cuts a gap through the Palaeocene-Eocene Calcaire de Brie that forms the scarp slope of the "Falaise de l'Île de France". Just north of Hautvillers on the road northwest from Épernay towards Fismes there is a superb panorama viewpoint over the landform assemblage here (Fig. 9.6a) from the (forested) Palaeocene-Eocene Calcaire de Brie scarp tops, the periglacially mantled midslopes (vine cultivation), down to valley wetlands along the Marne valley, through which the River Marne (albeit locally canalised) has tortuous meanders.

Incised Meanders, Misfit Modern Rivers   Throughout eastern France the main rivers are incised below Neogene erosion surfaces that cut across the main relatively resistant Mesozoic rocks (particularly Cretaceous Chalk, Cretaceous sandstones and Jurassic limestones). The switch from low gradient non-incising rivers to incised valleys appears to have occurred with early-mid Pleistocene climatic deterioration (see above and Chaps. 3 and 4). The main incision phases seemed to have occurred under high runoff conditions during the Pleistocene cold phases, resulting in large valley meanders. During the intervening interglacials (as now) river flows were much less and the morphology of small river meanders (misfits) within large valley meanders developed. There are numerous places in eastern France where such morphologies can be seen. Some of these sites are listed below.

The River Meurthe near Saint-Clément between Baccarat and Lunéville: Not really misfit meanders as such, the valley is a broad open valley slightly incised into Triassic and Lower Jurassic terrain, but the modern channel does have small tortuous meanders. Note that in this area there is a Quaternary incisional river terrace sequence.

The upper Moselle River between Chamagne and Bayon: Not really misfit meanders as such, the valley is a broad open valley of very large bends slightly incised into Triassic and Lower Jurassic terrain. The modern channel has well developed meanders of smaller geometry than the valley bends.

The Moselle River downstream of Thionville: Meandering channel within a broad open valley.

The upper Meuse River near Valcouleurs: Tortuous valley meanders cut into Upper Jurassic limestone, tortuous misfit modern meanders.

The River Meuse at St Mihiel: Tortuous valley meanders cut into Upper Jurassic limestone, tortuous misfit modern meanders.

The River Meuse downstream of Verdun: Incised meanders cut into Upper Jurassic limestones, tortuous modern misfit meanders.

The River Meuse south of Mouzon: Big double bends cut into Upper Jurassic limestone, modern channel misfit.

The Barr south of Charleville-Mézières: Huge incised valley meanders, within which the modern channel meanders are markedly misfit.

The Somme downstream of Peronne: Large incised valley meanders cut into the Chalk within which the modern anastomosing and meandering channel is markedly misfit.

The Upper Oise downstream of Guise: Classic incised meanders cut into the Chalk plateau, misfit modern channel meanders.

The River Aisne near Asfeld: Incised meanders cut into Chalk, modern channel small misfit meanders.

The Armançon between Rougemont (downstream from Montbard) and Lézinnes (upstream of Tonnerre): Incised meanders cut into middle and upper Jurassic limestones, modern misfit meanders.

The Cure downstream of Avallon: The Cure cuts into the Jurassic limestone terrain that fringes the Morvan: well developed incised valley meanders, modern small misfit meanders.

The Yonne between Clamecy and Cravant: Cut into the Jurassic limestone terrain that fringes the Morvan are well developed incised valley meanders, modern small misfit meanders.

The Central Paris Basin Not many people come to Paris for the geomorphology! So it is not easy to select appropriate highlight sites either within Paris nor elsewhere in the inner Paris basin, partly because of the relatively high degree of urban development and partly because many similar features are either better developed or more easily accessible in neighbouring regions. Such sites would include for example the escarpment of the Calcaire de Brie, crowned by the City of Laon because the escarpment is better seen in Champagne (see above, the previous section of this Chapter). Similarly, incised meanders with or without misfit modern meanders are better seen elsewhere than within the Île de France. However, within the Île de France there are such features on the Oise near Compèigne, the Aisne near Soissons, the Marne between Château-Thierry and Meaux, and the Seine both upstream and downstream of Paris. Incised meanders are well developed in Champagne (see above, the previous section of this Chapter) and Normandy (see above, the first section of this Chapter). The one site that would have more than local significance, would be the site on the Loing near Montargis from which Tourenq and Pomerol (1995) sampled the sediments which demonstrated a Massif Central origin and therefore that the early Pleistocene (?) palaeo-Loire flowed into the Seine (see above). However, I cannot track down the precise location of the site, so cannot be sure that it is still accessible (but see Highlights to the last section of this Chapter, below). The one Highlight location I would recommend is within the city of Paris itself.

Montmartre, the View from Sacré-Coeur This is the only site within Paris that I would highlight in relation to the geomorphology of the setting of Paris. Paris is a flood-prone city—some of that vulnerability is due to its geomorphological setting. The valley floor comprises an extensive floodplain formed probably during the latest Pleistocene. There were two former channel belts (Fig. 9.9 map). The southern one through the Île de la Cité/Notre Dame area is utilised by the modern river channel. The northern one ran through the Bastille area past the foot of Montmartre hill to rejoin the southern one in the west of the city. It was an active channel in the early Holocene but has acted only as a floodway since then. The two are separated by a fragment of low terrace deposited prior to the deposition of the floodplain—probably during the late Pleistocene. This terrace has very little elevaton above the floodplain. It is the site of the centre of the city. The modern channel (confined by floodwalls probably since the Middle Ages) follows the southern floodplain belt. In times of flood it cannot take all the floodwaters of the Seine and there is (at least partial) spillage into the northern floodplain belt. Since 1900 there have been eight such floods, and with global warming they might be expected to increase in frequency in the future.

Highlights: The Middle Loire Valley *Loire-Loing Capture Site at Briare* It is not easy to follow the details of this capture on the ground, so these notes can give only general directions. North of Briare is a tongue of Plio-Pleistocene sediments that heads NNE away from the Loire towards a broad open gap at the head of the modern Loing valley near La Bussière (Fig. 9.10 map). Presumably these sediments are of the proto-Loire. On the west side of the Loire, west and northwest of Briare, are other Plio-Pleistocene gravels that head westwards into the Sologne depression to the south of the great bend of the Loire through Orléans (Fig. 9.10 map). These presumably relate to the earliest post-capture westward course of the Loire. Since that time (sometime in the early Pleistocene) the Loire has migrated laterally towards the north to form the great bend through Orléans (Fig. 9.10 map)

Braided Channels of the Middle River Loire The Loire is the longest river in France. It is one of the largest lowland rivers, but unlike most of the others its modern channel is little altered by direct human activity. It carries a large (sand and silt) sediment load, and as a result, for much of its course, downstream of Nevers (the confluence between the Allier and the upper Loire) it has a braided pattern.

There are numerous sites on the Loire where the range of channel patterns can be seen. Below I make suggestions for field visits.

Downstream of Cosne: Braids, but an earlier abandoned meander can be identified within the floodplain.

Between Sully and Châteauneuf: Big meanders, but with sandbars and some islands—on the meandering/braiding threshold.

Between Beaugency and Blois: Mostly a single thread channel, but with sand-bars and occasional islands—not far from the braiding threshold.

Between Chaumont and Amboise: The channel is still relatively narrow, but sandbars increase in size and frequency downstream, with the channel becoming fully braided at, and downstream from, Amboise.

Between Tours and Villandry: Braided channel with sandbars and islands—interesting to compare with the meandering channel of the parallel River Cher, less than a km to the south of the Loire.

Between Candes-Saint-Martin (Vienne confluence) and Saumur: A fully braided reach.

The Authion reach of the Loire (Saumur to Angers): A wide braided channel with an extraordinary pattern of sandbar braids on the river bed (see Fig. 9.12c map, see also Gautier and Grivel 2006).

From Angers downstream the Loire enters its Vendée reach: the Armorican Loire (see Chap. 6).

Chapter 10
Aquitaine

This is one of the most attractive and varied lowland regions of France, with a warm but gentle climate, and an extensive sea coast.

The Geology of Aquitaine (Fig. 10.1 Map)

This region is bounded to the north and east by Hercynian massifs (Vendée to the north and the Massif Central to the east). In the southeast the boundary with the Massif Central is faulted, but further north the Jurassic rocks of the basin margin rest unconformably on the Hercynian rocks of the massif. In the far southeast, Cenozoic rocks of the basin lap onto the basement rocks of the Montagne Noire and extend through the Carcassonne gate into Languedoc to the east (see Chap. 14). To the far south the basin is bounded by the thrust front of the Pyrenees (see Chap. 11). The Aquitaine region itself falls naturally into two halves, divided by the valley of the River Garonne (Fig. 10.1 map). To the north and northeast of the River Garonne there are bands of Jurassic, Cretaceous and Paleogene rocks in successive belts away from the Hercynian massif (the southwestern part of the Massif Central). The coast, to the north of the Gironde estuary is low lying, dominated by saltmarshes, but to the south is a dune coast (see Fig. 4.7c), the Landes coast. To the south of the Garonne the country is essentially the Pyrenean foreland of Miocene sediments, forming the Lannemezan plateau (a Neogene 'megafan' of Pyrenean detrital sediment—Figs. 10.1 map, 10.2a, see also Fig. 4.6b). This is followed (to the southwest, west and northwest of Bordeaux) by a belt of Plio-Quaternary marine sediments.

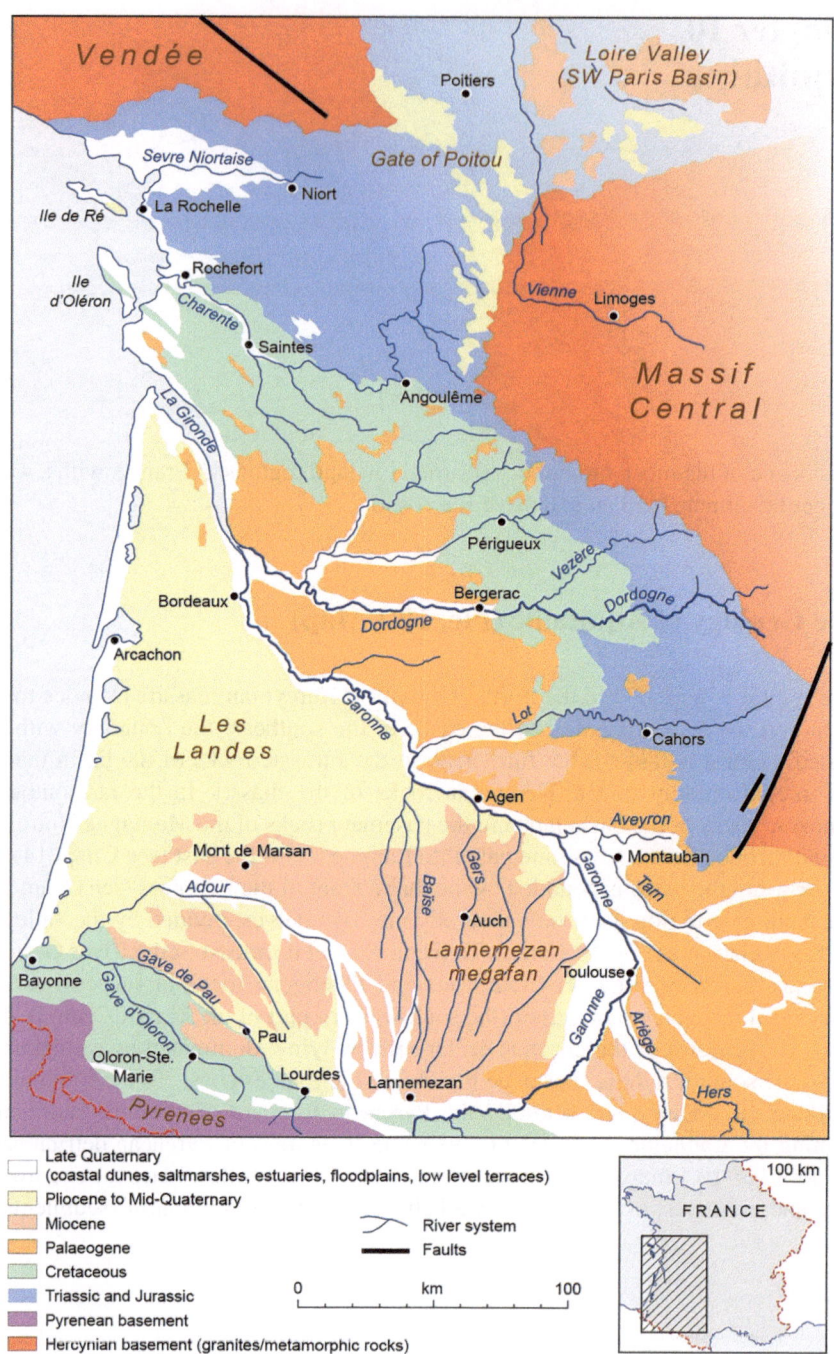

Fig. 10.1 Map of the Aquitaine basin

The Geomorphology of Aquitaine: Inland

The relief of the whole area of the Aquitaine basin is relatively subdued. The drainage comprises three main zones: (i) In the south several areas can be identified. The area west of the Lannemezan megafan is drained by rivers fed from the western Pyrenees (see Chap. 11). To the east of the Lannemezan megafan the River Garonne is fed from the central Pyrenees. To the east of the Garonne, apart from the Ariège (a tributary of the Garonne, fed also from the Pyrenees), all the main drainages are fed from the Massif Central (from south to north the main rivers are: the Tarn; the Aveyron; the Lot. (ii) In the east is an area of plateau surfaces sloping west from the western margins of the Massif Central, drained by rivers emanating from the Massif Central, especially the Dordogne and its tributaries. (iii) In the north is a lowland area drained in part by the Charente which is fed from the Limousin margins of the Massif Central (see Chap. 8) and in part by the diminutive Sèvre Niortaise which is fed primarily from the flanks of the Vendée (see Chap. 6).

West of the Garonne the landscape in the south of the Aquitaine basin is dominated by the Lannemezan megafan (see Chap. 4, Fig. 4.6b). This is an enormous Miocene-age alluvial fan which was fed with sediment from drainage basins in the then uplifting central Pyrenees. Its upper surface slopes away northwards (Fig. 10.2a). It is an enormous feature with a radius of around 100 km, falling in height from over 600 m at the apex to well below 200 m on its fringes. The main

Fig. 10.2 Southern Aquitaine, particularly the Garonne basin. (**a**) The surface of the Lannemezan megafan (see also Fig. 4.6b) looking south towards the Pyrenees from above Marciac. The concordant surfaces in the middle distance are dissected remnants of the megafan. (**b**) General view of the Garonne valley through Agenais with a wide weakly meandering river channel. (**c**). The enormous Gironde estuary

body of the megafan extends north from Lannemezan within the overall drainage of the Garonne, but there is a substantial portion that extends to the northwest within the current drainages of the Adour and the Gave de Pau. Since deposition and further uplift (during the late Neogene) it has been dissected by a radial pattern of drainage, bounded by the Adour on the west and the Garonne on the east. The radial rivers, primarily the Gers, which flows through Auch and joins the Garonne at Agen, and the Baïse which flows through Mirande and Condom (Armagnac country) to join the Garonne near Aguillon, are not particularly large rivers. They have small, locally tortuous meandering channels within narrow valleys cut into the megafan surface.

The upper Garonne, together with its main headwater tributary, the Ariège, is different. They are bigger, more dynamic rivers. They head in the Pyrenees (see Chap. 11). Within the foothill zone of the Pyrenees the Ariège and its eastern tributary the Hers, are transverse to a series of east-west thrusted fold structures (see Chap. 11). On leaving the Pyrenean foothill zone, the Ariège at Pamiers, and the Garonne at Cazères, join just upstream of Toulouse. Within Toulouse, the Garonne is joined from the southeast by a diminutive stream called the Ancien Hers (Fig. 10.1 map), the implication being that this was once the main course of the Hers. However, I can see no direct evidence to link this stream with the transverse course of the upper Hers described above, and can only assume that the association is purely a semantic one.

Beyond Toulouse the Garonne runs, with a mostly single-thread sinuous channel in a broad valley, through Agenais (Fig. 10.2b), through Moissac (Tarn confluence), through Agen, through Aiguillon (Lot confluence), eventually to Bordeaux and the western arm of the Gironde estuary (Figs. 10.2c, 10.3 map).

Downstream from Toulouse, within the Garonne system, there are two major east-bank tributary systems, the Tarn and the Lot (with their tributaries the Agout to the Tarn, and the Célé to the Lot). These rivers head in the Massif Central and cross the eastern part of the Aquitaine basin. They are deeply incised in their upper reaches within the southern part of the Massif Central (including the spectacular Gorges du Tarn see Chap. 8). Through the Jurassic limestones of the eastern part of the Aquitaine basin, they have incised meandering valleys including the spectacular incised meanders of the Lot between Cahors and Fumel (Fig. 10.4a, see also Fig. 1.2c). The lower reaches of the Garonne, through the Oligocene/Miocene terrain, tend to have relatively wide straight valleys within which the modern channels meander. The Garonne together with the Dordogne feed into the Gironde estuary, the largest estuary in France (Fig. 10.2c). These two rivers flank the "Entre-deux-Mers" peninsula—classic Bordeaux wine country (Fig. 10.3 map).

The Dordogne system includes a main tributary from the north, the Isle (with its tributaries the Dronne and the Auvézère), plus the direct tributaries to the Dordogne itself (the Cère and the Vézère). All these rivers head in the western part of the Massif Central (see Chap. 8), then cut across the Mesozoic then the Cenozoic rocks of the eastern margins of the Aquitaine Basin. Their courses across the Mesozoic rocks, here mostly Cretaceous rather than Jurassic limestones, are marked by incised meanders with the modern channels often misfit. Across the Cenozoic rocks further

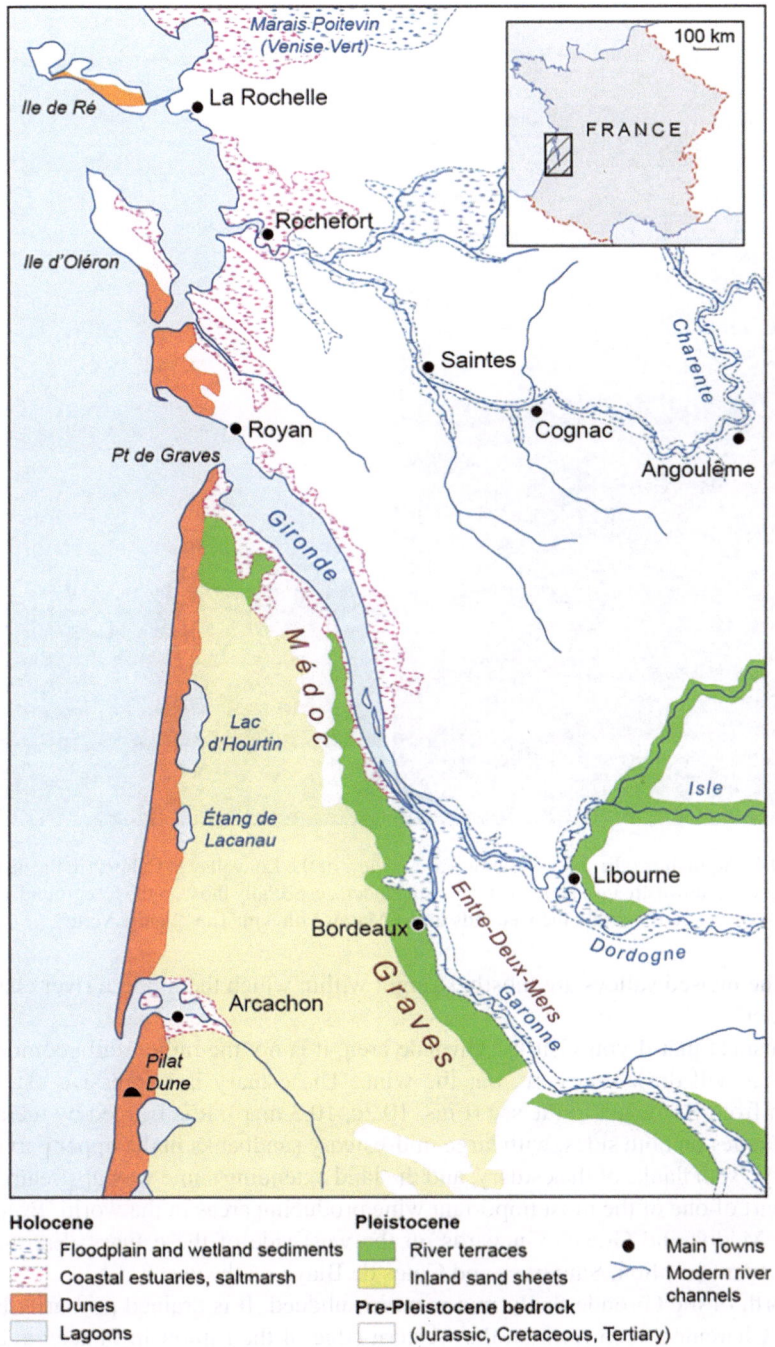

Fig. 10.3 Map of the Gironde estuary

Fig. 10.4 Aquitaine: other river systems. (**a**) Air view of the Lot valley at Cahors (in the distance), looking west (downstream). Note the incised meanders, especially those in the foreground. (**b**) The River Sèvre Niortaise within the wetlands of the Marais Poitevin: The "Venise Verte"

west the incised valleys are mostly straight within which the modern river channels meander.

I suspect that if you visit the Gironde area, it is not the rather dull geomorphology that will draw you here, but the wine. The estuary is impressive extending 60 km from Bordeaux to the sea (Figs. 10.2c, 10.3 map). It is fringed by reclaimed saltmarshes on both sides, with large mid-estuary sandbanks in the upper part of the estuary. Both flanks of the estuary, and the land extending some way upstream, form the heart of one of the most important wine producing areas in the world: Bordeaux itself; Médoc and Graves vineyards on the west side of the estuary; Entre-deux-Mers, Saint-Émilion, Sauternes and Côtes de Blaye on the east.

North of the Gironde the terrain is more subdued. It is drained primarily by the River Charente which rises on the western edge of the Limousin Plateau at elevations of barely 200 m. During the Early Pleistocene (?) the Charente lost its original headwaters to the Loire-draining Vienne (see Chap. 8). The Charente flows south through Angoulême, Cognac and Saintes to reach its marine estuary at Rochefort. It is a slow moving (sluggish!) river within relatively low relief valley meanders, within which, especially north of Angoulême, the modern channel is anastomosing through wetlands. The only other river of any significance is the Sèvre Niortaise. It is another low gradient low energy lowland river, which rises in the lowland area to the east of the Vendée massif, flows through Niort into the wetlands of the "Venise Verte" at Damvix (Fig. 10.4b), and then into the extensive Marais Poitevin marshes to reach the sea just north of La Rochelle.

The Aquitaine Coasts

There are major contrasts in coastal geomorphology north and south of the Gironde estuary. North of the Gironde (Fig. 10.1 map) is a lowland, complex, marshy coast. There are though, local low cliffs and rock platforms in Jurassic limestone at Jard in the far north of the area. Otherwise this section of coast is dominated by marshlands including the large Marais Poitevin, extending some 30 km inland almost to Niort. This was the former estuary of the Sèvre Niortaise. The river still runs through the reclaimed marshland. South of La Rochelle and extending south of Rochefort is another reclaimed marshland area, including the Charente estuary. The coastal fringes of both marshland areas are mudbanks and active saltmarsh. Offshore are two ("Tourist") islands, the Île de Ré and the Île d'Oléron, both now accessible via modern road bridges. The seaward sides of both islands have sandy beaches, but the sheltered landward sides, especially of the Île d'Oléron, have marshes and only small beaches. Inland of Marennes (on the mainland) is an extensive tidal marsh area, devoted to oyster farming. South from there, facing the open sea are dunes and sandy beaches, including spit features.

This northern coast contrasts with the coast south of the Pointe de la Grave, at the mouth of the Gironde estuary (Fig. 10.3). From there south virtually to the Spanish border, the coast is straight, fronted by a sandy beach and backed by dunes (see Fig. 4.7c), including the highest dune in Europe, at Pilat, just south of Arcachon. The surf is supposed to be good! Inland from the dunes themselves is a depression with a number of lagoons, beyond which is the flat (mostly plantation conifer-forested) Landes plain. It is a sand sheet blown inland from the dunes. Midway along this section of coast is a breach in the dunes, the Bassin d'Arcachon protected from the northwest by a long sand spit. The dunes however are only semi-natural, processes and morphology being influenced by management style (Bossard and Lerma 2020). Southwards, the dunes end at Bayonne, on the estuary of the Adour, beyond which is the cliffed coast that is essentially where the Pyrenees reach the sea. This cliffed coast continues to Hendaye (for about 15 km) on the Spanish border.

Highlights of Aquitaine: Inland

The Lannemezan Megafan The major plate-tectonic structures of the Pyrenees were emplaced during the Palaeogene (see Chaps. 2 and 3). Into the "mid-Tertiary" (the Miocene) there then followed a period of sustained uplift causing substantial sediment generation. North of the Pyrenees this resulted in the deposition of a huge megafan radiating northwards from a double apex, the main apex at Lannemezan and a secondary apex to the west of Tarbes. The huge scale of the megafan can only really be appreciated from satellite imagery (see Fig. 4.6b). At present the elevations of the megafan surface are at around 600 m in the apex areas, and at around 200 m in the outer (Armagnac) limits. The fan surfaces are drained by a series of mostly relatively shallow radial valleys incised into the surfaces. The southwest margin of the megafan complex is bounded by the Gave de Pau drainage (see Chap. 11). The two parts of the megafan complex are separated by the Adour and Arros drainages. The southeast margin of the megafan complex is bounded by the Neste/Garonne system. These river systems rise in the Pyrenees. All other drainages rise within the megafan complex itself and are radial in pattern, therefore receive few if any tributaries. All have small modern meandering channels. Only the Baïse and the Gers, both draining the main part of the megafan complex, could be regarded as even moderate sized rivers. To visit the megafan complex, I would recommend following one of the radial valleys up to Lannemazan, itself situated at the apex of the main (eastern part of) the megafan complex. From a point about 2 km south of Lannemezan there is a viewpoint over the incised Neste drainage which dissects the southeast corner of the megafan complex and there are views towards the Pyrenees (Fig. 10.2a).

Incising Rivers in the Eastern Part of the Aquitaine Basin These river systems rise mostly in the Massif Central to the east and then dissect the belt of Mesozoic (Jurassic and Cretaceous) limestone terrain, most with well developed incised meanders. All, except the Charente in the north, then cross the Cenozoic (mostly sand and clay) terrain in the basin centre, a terrain of lower interfluves, wide river valleys, often with Quaternary terraces and modern meandering channels but no incised meanders. The major systems are listed below (from south to north).

The Garonne System The main headwaters of the Garonne, both the Garonne itself and the Ariège are sourced in the Pyrenees (see Chap. 11). Downstream from the Pyrenean mountain front both rivers have wide valleys, sinuous to meandering channels, with the modern floodplain flanked in many places by Pleistocene terraces. Between Saint Gaudens and Cazères the Garonne is regulated by dams. The two rivers (Garonne and Ariège) converge at Toulouse. From there downstream the Garonne has a meandering channel but locally appears to be near the braiding threshold with gravel shoals and occasional islands. It is set in a wide floodplain, which near Grenade has traces of abandoned meanders. The floodplain is flanked by Quaternary terraces. Similar morphology persists beyond the Tarn confluence at Moissac, past the Lot confluence at Aiguillon, all the way to the tidal limit upstream of Bordeaux. Locally river levels are controlled by wiers. The main tributaries are right bank tributaries from the east, the Tarn (confluence at Moissac), and further north the Lot (confluence near Aiguillon).

The Tarn/Aveyron System The main east-bank tributary to the Garonne is the Tarn (confluence at Moissac), with its own tributaries, the Aveyron (north-bank) and the Agout (south-bank). All three head in the southern part of the Massif Central/Cévennes area. The Tarn heads in the northern Cévennes, south of Mont Lozère, crosses the karstic Causses plateau via the spectacular Gorges du Tarn (see Chap. 8), then traverses the complex basement of Rouergue (gorges, incised meanders see Chap. 8). Then it passes through the spectacular incised meanders at Ambialet (see Fig. 4.5c map) before reaching the Cenozoic rocks of the Aquitaine basin near Albi. Downstream from Albi the Tarn has a rather dull course, within a broad open valley via St Sulpice/la Pointe and Montauban to the Garonne. The channel itself in this area is sinuous to meandering but manipulated by a series of weirs.

At St Sulpice the Tarn is joined from the south by a left-bank tributary, the Agout. The Agout rises in the Languedoc/Montagne Noire Massif with incised valleys, locally incised meanders. Then, between Mazamet and Castres it enters the Cenozoic Aquitaine basin. From there it has a mildly sinuous to irregular meandering channel through a broad open valley, ultimately to join the Tarn at St Sulpice.

The Aveyron, a right-bank tributary of the Tarn, is an exciting river. Its headwaters are on the Causse de Séverac from where it clips the Causse Comtal, north of Rodez. It is then incised across the basement rocks of Rouergue (incised meanders) to Villefranche-de-Rouergue. There it crosses the Massif-Central bounding-fault and turns abruptly south to follow the fault, first of all in a fairly open straight fault-aligned valley, then in a series of spectacular fault-aligned incised meanders through Najac to Laguépie. At Laguépie it is joined by a left-bank tributary, the Viaur, which rises within and then crosses the Rouergue Massif through multiple spectacular incised meanders. Incidentally, along this valley, just east of Pampelonne is the Viaduc du Viaur, the well known steel-arch railway viaduct. The Aveyron then crosses through Jurassic rocks through the Gorges de l'Aveyron before passing onto the Cenozoic rocks of the Aquitaine basin, ultimately to join the Tarn just north of Montauban.

The Lot System Further north, the Lot is a major tributary of the Garonne (confluence at Aiguillon). Both it and its main north-bank tributary, the Célé, head in the Massif Central (see Chap. 8). Both then enter the Aquitaine basin, the Lot at Capdenac and at the Célé at Figeac, where Jurassic limestones are unconformable on the basement rocks of the Massif Central. The Célé, within the Aquitaine basin is mildly incised with a floodplain. Towards the Lot confluence at Bouziès are narrow and incised reaches. Incidentally, this is near the well known medieval village site of St Cirq-Lapopie, and particularly the archaeologically-rich cave of the Grotte du Peche Merle, famous for its late Pleistocene/Upper Palaeolithic cave paintings, dating from between 27 ka BP and 18 ka BP (for context see Chap. 3). On the Lot, upstream from Cahors, there are incised migratory meanders with Pleistocene terrace remnants preserved in places. Downstream from Cahors to Fumel there are perhaps what are the most spectacular regular incised meanders in France (Fig. 10.4a, see also Fig. 1.2c). The channel is migratory within the floodplain and there is local preservation of terraces. At Fumel the valley widens, is flanked by Cenozoic rocks of the Aquitaine basin, and is straighter. The channel is more irregular, sinuous to locally meandering to the confluence with the Garonne at Aiguillon.

The Dordogne System   The Dordogne emerges from its upper reaches in the Massif Central (see Chap. 8), crosses the bounding fault near Beaulieu-sur-Dordogne into a relatively wide open valley containing river terraces. It has a sinuous modern channel. Near Martel it enters a meandering valley, within which there are terraces and a modern floodplain, an indication of a long (Pleistocene) period of incision into Jurassic limestones, accompanied by channel migration. This type of morphology, but cut into Cretaceous rocks, persists downstream through the Sarlat area to about Lalinde (several km upstream of Bergerac). From here on, the valley is wide, more or less straight and cut into the basinal Cenozoic rocks. The modern meandering channel is flanked by floodplain plus Quaternary terrace fragments. The north bank tributaries of the Dordogne (Îsle, Dronne, Auvézère, Vézère) have similar morphologies; laterally migrating single channels within incised meanders in the upstream reaches through the Cretaceous limestones, wider straight valleys through the Cenozoic rocks of the basin, within which the modern channel laterally migrates by meandering.

The Charente   Beheaded by the Vienne (see Chap. 8), the upper Charente above Chatain is a tiny meandering stream in a broad, little-incised valley, manifestly underfit. Only downstream from Civray does the depth of incision into the Jurassic limestone plateau, although not great, become noticeable. There it has shallow incised meanders, within which the modern channel is manifestly underfit. This morphology continues through Ruffec to Mansle. From there to Angoulême the modern misfit channel has a complex anabranching pattern rather than a simple meandering one. Beyond Angoulême the valley runs northwest following the Upper Jurassic/Cretaceous boundary and at the foot of the escarpment of the Upper Cretaceous (chalky) limestone. The incised meanders then become less obvious so that by Jarnac then Cognac (important vine-growing area) and Saintes the wide valley is little incised and the modern channel is sinuous to meandering.

Highlights of Aquitaine: The Coast

The Aquitaine coast is very much a lowland depositional coast, but with enormous contrasts north and south of the Gironde estuary, the largest and most impressive tidal estuary in France. To the north saltmarshes dominate the Charente coast; to the south dunes and lagoons dominate the Landes coast.

The Charente Coast   Salt marshes dominate the Charente coast. At l'Aiguillon-sur-Mer there is a depositional spit with recurved lateral ridges that indicate a north to south drift. Behind the spit are creeks and saltmarshes. A little to the south and east is the Anse d'Aiguillon bay with saltmarshes fringed by mudflats. Inland is the Marais Poitevin which includes the meandering channel of the Sèvre Niortaise running through the now mostly reclaimed marshland, the "Venise Verte" area near Damvix (see Fig. 10.4b). Further south is the Charente estuary and Châtelaillon bay, another area of mudflats plus reclaimed saltmarsh.

The Landes Coast   This is a dune coast (see Fig. 4.7c, see also Bossard and Lerma 2020) with lagoons inland. I recommend several areas.

Hourtin Plage: Coastal sand beaches that are backed by mostly active dunes. Inland is the brackish Lac d'Hourtin-Carcans.

Bassin D'Arcachon (See also Bertrand, Chap. 8 in Fort and André, Eds. 2014): The one breach of the dunes in the whole of the Landes coast. This is a tidal embayment with mudflats (these are used for oyster cultivation), saltmarsh, and to the south, offshore sand bars.

Pilat: South of the Bassin d'Arcachon the dune system resumes, including the highest dune in Europe, Pilat (see Fig. 4.7c).

Chapter 11
The Pyrenees

The Alpine Regions

So far we have looked at what could be described geomorphically as the least inter-esting regions of France. We now turn to regions which could be described from the viewpoint of geomorphology as the most interesting, involving: (1) Cenozoic tec-tonics, creating the Pyrenean and Alpine mountain chains and adjacent areas; (2) Quaternary (and modern) mountain glaciation; and (3) active modern processes.

The plate-tectonics context of the Pyrenean/Alpine mountain systems as a whole has already been described (Chaps. 2 and 3). This context involved: the Palaeogene rotation of Iberia resulting in the development of the Pyrenees, and associated struc-tures; and the complex collision of the African and European plates culminating in the Miocene, resulting in the forward thrusting of the French Alps and Jura, and the associated faulting. The broad effects of Quaternary (and modern) glaciation in the Alpine regions have also been described earlier (see Chap. 3, and the appropriate sections of Chap. 4).

The Alpine region as a whole is an enormous and exciting region. The term "Alpine" as it is used here is a *geological* term that relates to the Palaeogene rotation of Iberia, and the "mid-Tertiary" plate-tectonic setting of Africa encroaching on the southern margin of Europe (see Chaps. 2 and 3), creating major mountain systems throughout southern Europe. Within France this includes the Pyrenean range (Chap. 11), plus the whole of the French Alpine ranges, extending from the Jura mountains in the north (Chap. 12), through the "classic" northern Alps (Chap. 13), south into the Provençal Alps, to the Mediterranean coast (Chap. 14). Within Chaps. 12–14 I also deal with the associated lowland areas.The Alpine/Pyrenean region also includes the island of Corsica (Chap. 15).

These "Alpine" chapters are as follows:

Chapter 11: The Pyrenees
Chapter 12: The Jura Mountains and the Saône Basin

© The Author(s), under exclusive license to Springer Nature
Switzerland AG 2025
A. Harvey, *The Geomorphology of French Landscapes*,
https://doi.org/10.1007/978-3-031-68490-6_11

The Geology of the Pyrenees

The Pyrenean zone was created in late Cretaceous to Eocene time, by the rotation of
the Iberian platelet from its previous location in the Bay of Biscay area, then its
northward movement and collision with the European plate. The main structural
deformation occurred during the Eocene forming a narrow west to east mountain
belt, the Pyrenees, along the Franco-Spanish border (Fig. 11.1 map, see also Chap.
2, and Figs. 2.4, 3.1 map). The mountain range continues westwards, though at
more subdued elevations, along the north coast of Spain as the Cantabrian Mountains.
Although the main structural deformation took place during the Eocene, major post-
orogenic uplift continued well into the Miocene.

The core zone of the Pyrenees (Fig. 11.1 map) comprises granites and metamorphic rocks, mostly of re-incorporated Hercynian rock units. The rocks have been
thrust from the south resulting in a marked west to east terrain alignment. To the
north of the core mountain zone, both in the west (the Basque country) and especially in the east, are west to east folded Jurassic and Cretaceous sedimentary rocks
of the Pyrenean foreland (Fig. 11.2 map). These structures add to the strong west to
east linearity of the terrain. Across the Carcassonne gate, the Montagne Noire (of
granites and metamorphic rocks) is essentially a detached block of Pyrenean structures, thrust against the southern margin of the Massif Central. Further east still,
Pyrenean rocks and structures continue into southern Provence (see Chap. 14).

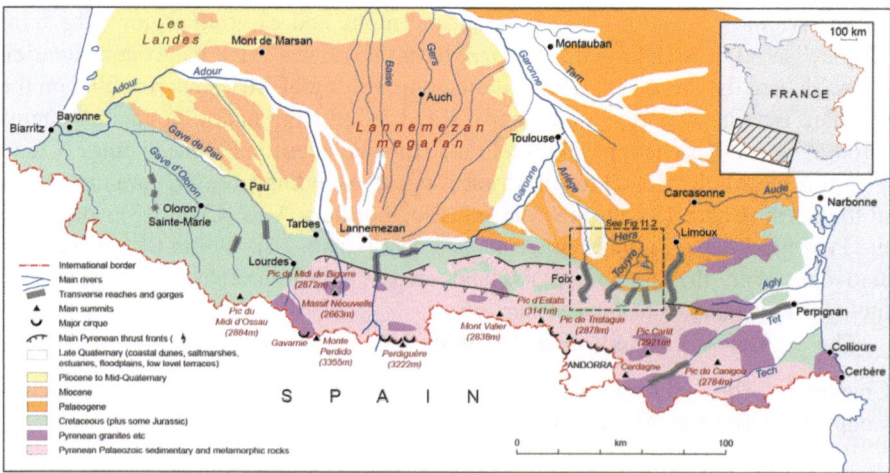

Fig. 11.1 Map of the Pyrenees

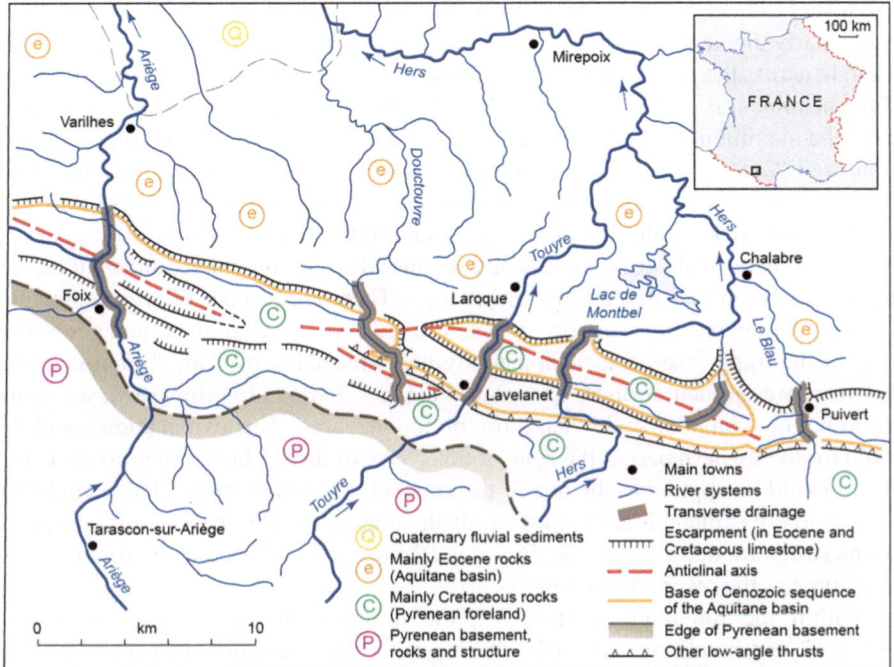

Fig. 11.2 Map of the transverse drainage in the northern foothills of the Pyrenees

The Geomorphology of the Pyrenees

The high relief of the axial zone of the Pyrenees is due primarily to "mid-Tertiary" post-orogenic uplift. The present elevations of the peaks range between 600 and 1400 m in the western Pyrenees (west of the Larrau Pass), but rise to over 3000 m in the axial zone of the central Pyrenees (between the Larrau Pass and the Andorra border) then decline through the eastern Pyrenees to barely 700 m near the Mediterranean coast.

Continued tectonic activity during the Palaeogene, involving compression from the south, folded the Cretaceous limestones of the foreland zone into a series of west to east aligned anticlines and synclines (Fig. 11.2 map). During the Miocene enormous amounts of sediment were generated in the high central Pyrenees. These were carried on the French side of the divide by the palaeo drainages of the Garonne / Neste, Adour and Gave de Pau into the Lannemezan area to form the Lannemezan megafan complex (see Chaps. 3 and 10, see also Fig. 4.6b). This megafan buries the Pyrenean foreland structures.

East and west of the Lannemezan megafan complex the Pyrenean foreland structures are exposed. These structures are a series of west-east orientated anticlines and synclines in Cretaceous and Palaeogene rocks which form a series of west-east orientated hill ridges. A curious aspect of the geomorphology of both these zones, particularly that in the east (Fig. 11.2 map), is that the drainage is transverse, cutting straight across the ridges and their fold structures. There are two possible origins for this phenomenon. We can rule out glacial influences; these zones are far beyond even the maximum Pyrenean glacial limits. The two possibilities are "superimposition" and "antecedence". Superimposition would need a cover rock to have been deposited unconformably over the fold structures. The drainage that was initiated on the cover rock would then have cut through into the underlying fold structures. The obvious possible cover rock would be the Miocene fluvial sediments those in the Lannemezan megafan a little further west. There are patches of such sediments capping the divides between the Arriège, Lèze, Arize and Garonne drainages to the NW of the transverse courses, apparently unconformable over the fold structures. Furthermore on their surfaces are small drainages divergent northwestwards, similar to but on a smaller scale than those on the megafan. This explanation could be valid for the western part of the zone of transverse drainage, but there is no evidence that it could be applied to the rest of the zone. On the whole antecedence might be a more likely explanation, in other words the drainage would have originated prior to the structural deformation (ie. Pre-mid(?)Miocene), and become incised into the structures as they were developing.

Within the mountains themselves, the drainage divide (between northerly "French" and southerly "Spanish" drainage) mostly coincides with the ridges of high peaks, themselves coincident with the major nappe structures. There is however, one major exception and that is the head of the Garonne system. This river system is sourced within the Vall d'Aran on the Spanish side of the main ridge, cutting across the ridge into France through a structurally-controlled gorge southeast of the small village of Fos. Is this another case of "antecedence"?

One of the major influences on the relief and the geomorphology of the Pyrenees is Pleistocene glaciation. The glacial extent in the Pyrenees is today, and was in the Pleistocene, much less than that in the Alps. The maximum extent of glaciation during the penultimate glaciation is not clear, but probably exceeded that of the last glaciation. Ice extent during the last glaciation just about reached the mountain front of the high Pyrenees. For example there is a series of last glacial terminal moraines, together with the remnants of an ice-marginal lake near Lourdes, both dating from about 34 ka BP (somewhat earlier than the last glacial maximum in the Alps). They were deposited at the margin of a complex valley glacier that came down the valley of the Gave de Pau. The other major valleys of the central Pyrenees would have been similar. These long valley glaciers rapidly melted back, so that by the late Pleistocene and into the Holocene, glaciation was restricted to a number of small cirque glaciers and ice patches in three areas. These areas are: around the Vignemale Massif in the west central Pyrenees near Cauterets; further east in the Pic Perdiguère area; and in the Tristagne range on the Andorra border. The glacial extent was less in the eastern Pyrenees (Delmas 2005). The only modern glacier within the Pyrenees "larger than a cirque glacier" is the Glacier d'Ossoue in the Vignemale Massif. Today, all the Pyrenean glaciers are rapidly melting back. It is predicted that by the year 2050 there may be none left!

Fig. 11.3 The western Pyrenees (See also Fig. 3.7a cirque, Néouvielle massif). (**a**) Pic d'Aubert, Néouvielle area: At this high elevation there are snow patches even in high summer. There are also small glaciers in this area. Note also the severe frost weathering of the pinnacled ridge. (**b**) High on the main Pyrenean watershed at the Col du Pourtalet: View north to the Pic du Midi d'Ossau. In the foreground is dissected Quaternary morainic country. (**c**) Lac d'Orédon, Néouvielle Massif: a modern fan delta building out into the lake. Note also the scree slopes above the lake

The Pleistocene glacial legacy is expressed in a wealth of spectacular glacial erosional landforms (Fig. 11.3a): arêtes; cirques (see Fig. 3.7a the Col d' Aubert, S of the Néouvielle Range); troughs; hanging valleys etc. Perhaps the most spectacular area is the Gavernie/Troumouse area at the head of the Gave de Pau valley—cirques, waterfalls, glacial erosional landforms (see Fig. 1.3b). Not far away is the valley of the Gave d'Ossau, which includes the roads up to the Cols d'Aubisque and Pourtalet, both of which give spectacular views of glacially eroded and glacial depositional terrain (Fig. 11.3b). Beyond the immediate glacial limits Pleistocene periglacial processes were active. Throughout the high mountain country, there is evidence, as would be expected, of very active modern geomorphic processes—active scree slopes, hillslope gullies (even some badland areas), gullies feeding debris-flow lobes, debris cones, alluvial fans, fan deltas (Fig. 11.3c) and gravel-bed braided river channels. All such features are particularly well developed along the road to the Bielsa tunnel (Fig. 11.4a, b).

The eastern Pyrenees differ in a number of ways from the high central Pyrenees. Some of these differences are structural. Although the structural geology is dominated, as throughout the Pyrenees, by the northward thrust sheets of metamorphic rocks and granites, here there are also important extensional faults. These faults relate to the 'mid-Tertiary' plate-tectonic extension in the western Mediterranean. This extension led to the formation of several extensional sedimentary basins within

Fig. 11.4 The central Pyrenees (See also Fig. 1.3b Cirque de Gavarnie). (**a**) The Bielsa valley, central Pyrenees: Rock slopes feeding partially vegetated screes, also gullies and gullied fans. Note that in this valley are numerous indications of active modern processes. (**b**) A small alluvial fan in the Bielsa valley, central Pyrenees: Note that the apex area of the fan is lightly trenched, probably following Late Pleistocene sedimentation. Note also that most of the fan surface is vegetated and recent sediments are restricted to the distal area

the eastern Pyrenees, the Cerdagne, Capcir and Conflent basins and the Rousillon basin on the coast (see last section of Chap. 14).

In addition, major southwest to northeast faults developed along which the Têt and Tech drainages are aligned. A second contrast between the central and the eastern Pyrenees relates to Pleistocene glaciation. Apart from isolated cirque glaciers within the Carnigou range, Pleistocene mountain glaciers got no further east than the flanks of the Capcir basin to the east of the Carlit range (Delmas 2005).

Thirdly, the modern climate of the eastern Pyrenees is distinctly Mediterranean, characterised by summer drought but high rain intensity during (especially autumnal) storms. Therefore the geomorphology is more a reflection of modern fluvial processes in a tectonically-active landscape than of Pleistocene glaciation. Distinctive are the gorges of the upper River Têt and badlands cut in Pliocene silts, near Ille-sur-Têt (For more details of the Mediterranean drainage, the Aude and the Têt systems, see end of Chap. 14). Finally, the Pyrenees end on a short stretch of spectacular cliffed Mediterranean coast (see also end of Chap. 14).

Interestingly, the eastern Pyrenees underwent two periods of major plate-tectonic activity. The first was the main Pyrenean mountain-building period in the Late Cretaceous to the early Eocene. This was followed by a substantial period of post-tectonic uplift in the Palaeogene, then in the "Mid-Tertiary" by a period of quiescence, during which time extensive peneplain surfaces developed. These are still evident within the high mountains of the Eastern Pyrenees. The second phase of tectonic activity occurred in the (late?)Miocene, causing km-scale uplift of the Carnigou massif (south of the Têt valley), which initiated the present wave of deep dissection (see below).

Highlights of the Pyrenees

Introduction To get an overview of the geomorphology of this spectacular mountain area I recommend a **one-day tour** within the northern part of the central Pyrenees. This is a route that skirts the high country and would give a good one-day overview of the western and central Pyrenees, with exceptional mountain views into the high Pyrenees. Travelling from west to east, the route starts south of Pau on the D934 towards Laruns.

On the D934 drive south through the village of Louvie-Juzon (just south of Arudy/Izeste) and into the high mountains along the valley of the upper Gave d'Ossau. At Laruns turn east onto the D918 towards the **Col d'Aubisque.** En route to the col note the bedrock-incised stream channel with waterfall, and the hillslope gullies above. At the col (1709 m) there are spectacular views of the high Pyrenees to the south, including the Pic de Ger (2613 m). From this col continue on the D918 into the valley of the Gave d'Arrens at Arrens-Marsous. There the D918 turns left down the valley to Argelès-Gazost in the much larger valley of the Gave de Pau. In Argelès-Gazost cross the town and the river, and turn right (south) on the D913/921 towards Luz-St Sauveur. In the first km or so note the braided channel of the river, the Gave de Pau. At the roundabout after Pierrefitte-Nestalas, make sure you turn left (east) towards Luz. For much of the way the road is in the steep sided valley of the Gave de Pau, the river itself in an incised channel. In the centre of Luz turn left (east) onto the D918 towards the Col de Tourmalet and Bagnères-de-Bigorre.

On the way up to the **Col de Tourmalet** (2118 m) note the deeply dissected northern valley-side slopes in the vicinity of the hairpins that lead to the final climb to the col. The views from the col itself are stunning, especially those to the south to the flanks of the Néouvielle Massif with Pleistocene glacial features very much in evidence. On the way down from the col note the waterfall, the Cascade du Garet (on the right of the very large hairpin bend to the south). From there the road follows the valley bottom to Ste-Marie-de-Campan. Here turn sharp right (still on the D918) towards Arreau.

En route towards Arreau the road climbs, initially through wooded terrain, then open country, to the **Col d'Aspin** (1489 m), less dramatic than the previous cols, but nevertheless with extensive views north and south over the central Pyrenees. Continue down from the col into the Neste valley at Arreau. The Neste is one of the main headstreams of the Garonne, and in the Neogene was one of the major source zones for the Pyrenean frontal Lannemezan megafan (see Chap. 10). At Arreau turn left (east) onto the D618 heading for Bagnères-de-Luchon. You are now following the Lauron valley, tributary to the Neste, and from Avajan turn left, climbing above the east side of the valley. Beyond Loudervielle and the Mont hairpins the road climbs towards the **Col de Peyresourde** (1569 m). Like the Col d'Aspin this col is not particularly spectacular, but it has distant views of the high Pyrenees beyond Luchon. From the col the road follows the Larboust valley (of the River One) down to Bagnères- de-Luchon.

From Bagnères-de-Luchon there are two possible routes, one running through a corner of Spain, the other slightly longer, wholly within France. For the shorter route crossing temporarily into Spain, head south out of Bagnères then almost immediately turn left (still on the D618), passing St Mamet and heading for the **Portillon de Burbe** (1298 m), the col on the Spanish border. This col again has local mountain views without being particularly spectacular. The road then descends through numerous hairpins down into the upper Garonne valley at Bosost (still in Spain). Follow the (Spanish) N230 northwards downstream along the Garonne valley to the French border, at a point where the Garonne valley breaches a major west-east Pyrenean ridge. I do not know the details of the geology/geomorphology of this site. Continue north along the main road (the French N125) following the Garonne downstream to St-Béat, just short of where the Garonne makes another breach through a west-east Pyrenean ridge. (This is where the route through Spain rejoins the wholly French itinerary). For the wholly French itinerary, from Bagnères drive north along the D27/D125 down the valley of the Pique to the roundabout for Cierp-Gaud/Marignac, where you turn right (east) onto the D44 to St-Béat to rejoin the alternative route.

From St-Béat follow the D44 east to the **Col de Menté** (1343 m). In areas such as this, one wonders about the long-term environmental impact of ski resorts. From here the road descends to the D85 in the Ger valley. Turn left downvalley to the D618 junction. There turn right onto the D618 up to the **Col de Portet d'Aspet** (1069 m). There are views of the frontal Pyrenean ranges. From there the road descends to Saint-Lary then down the Bouigane valley to Audressein. There, take a left turn along the Lez valley to Saint-Girons. I suggest that this should be the end-point of this itinerary. You could continue east towards Foix. Alternatively you could return west by heading northwest out of Saint Girons and out of the Pyrenees.

The Western and Central Pyrenees   The Pyrenees include rugged high mountain landscapes, and for access (especially to the small residual glaciers) heavy hiking may be necessary. On the other hand there is road access into the high country, with many minor passes across the watershed into Spain.

Col du Pourtalet   A minor pass into Spain that gives good road access into the high country around the Pic du Midi d'Ossau (Fig. 11.3b). With some moderate hiking there is access to spectacular previously glaciated mountain landscapes in the Moines Massif, several kilometres northwest from the French side of the col.

Cirques de Gavarnie, Estaubé, and Troumouse (See also Fort, Chap. 12 in Fort and André, Eds. 2014) One of the real highlights of the Pyrenean mountain landscapes (see Fig. 1.3b) is this assemblage of three spectacular cirques. Their backwalls form the main Pyrenean frontier and divide. This mountain wall, especially on the west/Gavernie backwall, is also one of the main sites of modern glacial ice within the Pyrenees (hardly more than small ice patches, rather than glaciers of any size). Pleistocene glacial erosional forms of the cirques themselves provide the basis for spectacular Holocene forms (waterfalls, especially in Gavernie; a large alluvial fan, plus braided channels in Estaubé; waterfalls and incised bedrock channels at Troumouse). There is road access, via Luz and Gavernie, but a lot of tough hiking could also be involved.

Perdiguère Range, Bagnères-de-Luchon Area Excellent views of the Perdiguère range, from an easily accessible ski resort at Superbagnères (south of Bagnères-de-Luchon, accessible from the Lys valley, south of Bagnères). This range is the second major area within the Pyrenees of active modern glacial ice, albeit only in small ice patches, rather than major glaciers (Gavernie is the other area). Although Superbagnères is easily accessible, the Perdiguère range itself to the south is not, and would involve heavy hiking—be warned!

Pyrenean Front Ranges *Transverse drainages of the Ariège, Touyre and Hers Rivers* (see Fig. 11.2 map) To the north of the major structures of the Pyrenees that involve a wide range of (mostly crystalline) rocks, is a zone of northward thrust sheets. These involve mostly Mesozoic sedimentary rocks folded into tight east-west aligned synclines and anticlines. These form the Front Ranges of the Pyrenees. Bizarrely the main drainage emanating from the Pyrenees, particularly the Ariège and Hers river systems, have cut incised courses directly across these structures, either by superimposition or by antecedence (see above).

The Eastern Pyrenees The eastern Pyrenees differ in several ways from the western and central Pyrenees. They are slightly lower in elevation, therefore the extent of Pleistocene glacial ice was much less. Hence, other signals in the landscape are clearer, particularly the tectonic signal and the periglacial signal. In addition, the modern climate and vegetation have distinctly Mediterranean qualities. To see something of the geomorphology of this area, I suggest three separate itineraries that take in the major features of the area.

Itinerary 1: Upper Ariège Valley and the Col de Puymorens The upper Ariège valley, from Tarascon-sur-Ariège to Ax-les-Thermes and then south to the Puymorens and Puigcerdà areas, probably represents the last, eastward extent of the high Pyrenean (glacial) landscapes. From Tarascon, follow the N20 road to Ax-les-Thermes, following the floor of the Ariège valley. The Ariège has a single-thread mildly sinuous alluvial channel within a modern floodplain set below what are probably Late Pleistocene terraces. Opposite Verdun, at Seconec, is a large presumably Late Pleistocene tributary junction alluvial fan, fed by a northern tributary.

Beyond Ax-les-Thermes, continue south on the N20 still following the River Ariège upstream, but it is now mostly in an incised channel within a narrow valley. Note the debris cone fed from the east at Mérens-les-Vals; also the waterfalls. At L'Hospitalet-près-L'Andorra do not take the tunnel (the N20) but take the high road over the Col de Puymorens (the N320). From the col note the views of the spectacular Late Pleistocene intensely glaciated landscape visible towards the northwest. Descend southwards from the col (rejoining the N20) to Bourg-Madame in the Segre valley, opposite Puigcerdà on the Spanish border. The Segre River, is a "Spanish" river that from Puigcerdà heads west eventually towards the Ebro. To the east (within France) is the Carlit high plateau, a high-level erosion surface, which lies between the (Spanish) Segre drainage and the eastward drainage of the Tet (see below). It is an area that at least during the last glaciation escaped the direct effects of glacial ice. It is mantled by periglacial patterned ground. Further east is the Capcir basin, within which are the Last Glaciation terminal moraines of glaciers that flowed around the Carlit plateau. Apart from high level cirques in the Carnigou massif to the south of the Têt valley (see below), these moraines are the most easterly limits of any extensive Pleistocene Pyrenean glaciers. The Spanish border at Puigcerdà marks the end of this itinerary—from there either turn right (west) into Spain, or left (east) into the Têt valley.

Itinerary 2: The Têt Gorges (See also Calvet et al. Chap. 13 in Fort and André, Eds. 2014) This excursion could follow on from that described above. However, if that were to be the case, it would be followed in reverse sequence from the way it is described here. Here, it is described as a trip up the Tet valley from east to west, rather than the other way round. Incidentally, it is also the route followed by "Le petit train jaune", a major regional tourist attraction.

Start from Ille-sur-Têt, within the valley of the River Têt, at the westernmost point of the Rousillon lowland (see last section of Chap. 14). The lowland is a sedimentary basin formed of a stack of Pliocene to early Pleistocene fluvial sediments, forming a basin-filling sequence. Since the early Pleistocene the river has cut down, exposing these sediments on the margins of the basin (north of Ille) as dissected quasi-badland terrain. From Ille downstream much of the River Têt has an artificial channel, fed from the lake impounded by the Vinca dam (see below). Almost immediately upstream from Ille the channel is rock-cut, providing a site for the dam which retains the Vinca Reservoir. Above the reservoir the channel is an alluvial single-thread channel, set below the Late Pleistocene terraces which themselves are inset within the Neogene Conflent basin. This basin extends upstream as far as Prades. It is a half-graben bounded on the southern side by a major fault system, a system of regional faults extending from the coast inland to the head of the Têt drainage in the Cerdagne. South of the faults is the Carnigou massif (maximum elevation 2784 m) of basement rocks uplifted during the Neogene. This elevation was sufficient to sustain small cirque glaciers during the Last Glaciation.

Upstream from Prades the Têt is aligned along this fault system in a narrow valley, mostly with the channel incised into bedrock. At Olette the channel cuts from one fault to a parallel fault, the upstream segment to the south. For the first km or so upstream from Olette, the valley is tortuous within which the channel is incised into bedrock. Further upstream the valley floor is straighter, and broadly fault aligned. The north wall is deeply ravined into the bedrock with signs of modern active channelized debris flows, especially opposite Thuès-Entre-Valls, a village built on a (Pleistocene?) tributary debris cone. At Sauto, Gisclard suspension bridge carries the railway (with the "Petit train jaune") over the Têt Gorge. The gorge marks a major knick point in the valley profile where the river is incised below the Col de la Perche plateau. Above the gorge, beyond Mont-Louis, the river is only lightly incised into the eastern part of the Cerdagne basin. The main (central and western) part of the basin towards Puigcerdà is drained by the lightly incised headwaters of the Segre River (part of the Ebro system: see above).

Itinerary 3: The Spectacular Cliffed Coast at the Eastern (Mediterranean) End of the Pyrenees   From Perpignan head southeast on the N114 to Argeles and Collioure. From there south to Cerbère on the Spanish border, there is a short but attractive section of cliffed coast marking the seaward end of the Pyrenees. The cliffs are cut in Precambrian to Lower Palaeozoic metamorphic rocks of the Pyrenees. The small intervening bayhead beaches are not too much modified by urban development. Access is fairly easy at a number of points between Argelès and Cerbère.

Chapter 12
The Jura Mountains and the Saône Basin

Two very different regions are linked together here because their origins are linked: the Jura mountain chain, and to its west the Saône graben. This area as a whole is perhaps less well known than it could (should?) be. Perhaps this is because it is often by-passed by people travelling further south, either by TGV or on the motorway system. In several ways it is half way through France, and combines elements which are clearly "northern" with those which at least hint of the "south". In a trivial sense it is where the Michelin northern and southern map sheets meet! It is where, during the Second World War, occupied and Vichy France met. The Jura mountains are clearly "Alpine", but are much less spectacular than the Alps themselves. In a cultural/architectural sense it is where northern and southern elements meet. In the northern part of this region (eg. in Chalon-sur Saône) traditional buildings are clearly "northern" with steep roofs and flat tiles. In the southern part of the region (eg. in Mâcon), the traditional buildings are "southern" with low-angle roofs, "curled" so called provençal tiles. In Tournus, halfway between these two towns, the symbols are mixed. Above all, this is one of the great wine regions of France, as well as (incidentally) having interesting geomorphology.

The Geology of the Jura/Saône Region: (Figs. 12.1 Map and 12.2)

The Jura mountain chain is a chain of relatively low fold mountains aligned more or less northeast-southwest to north-south along the French/Swiss border (Fig. 12.1 map). It dominantly comprises folded Jurassic sedimentary rocks in three distinct units (Fig. 12.2). The most easterly unit, the internal Jura (the "High Jura"), lies along the Swiss border and forms the highest relief. It comprises mainly Middle Jurassic limestones in broad folds which are mostly expressed in the terrain as a resequent relief of anticlinal ridges and synclinal valleys (see Fig. 2.3 for

© The Author(s), under exclusive license to Springer Nature
Switzerland AG 2025
A. Harvey, *The Geomorphology of French Landscapes*,
https://doi.org/10.1007/978-3-031-68490-6_12

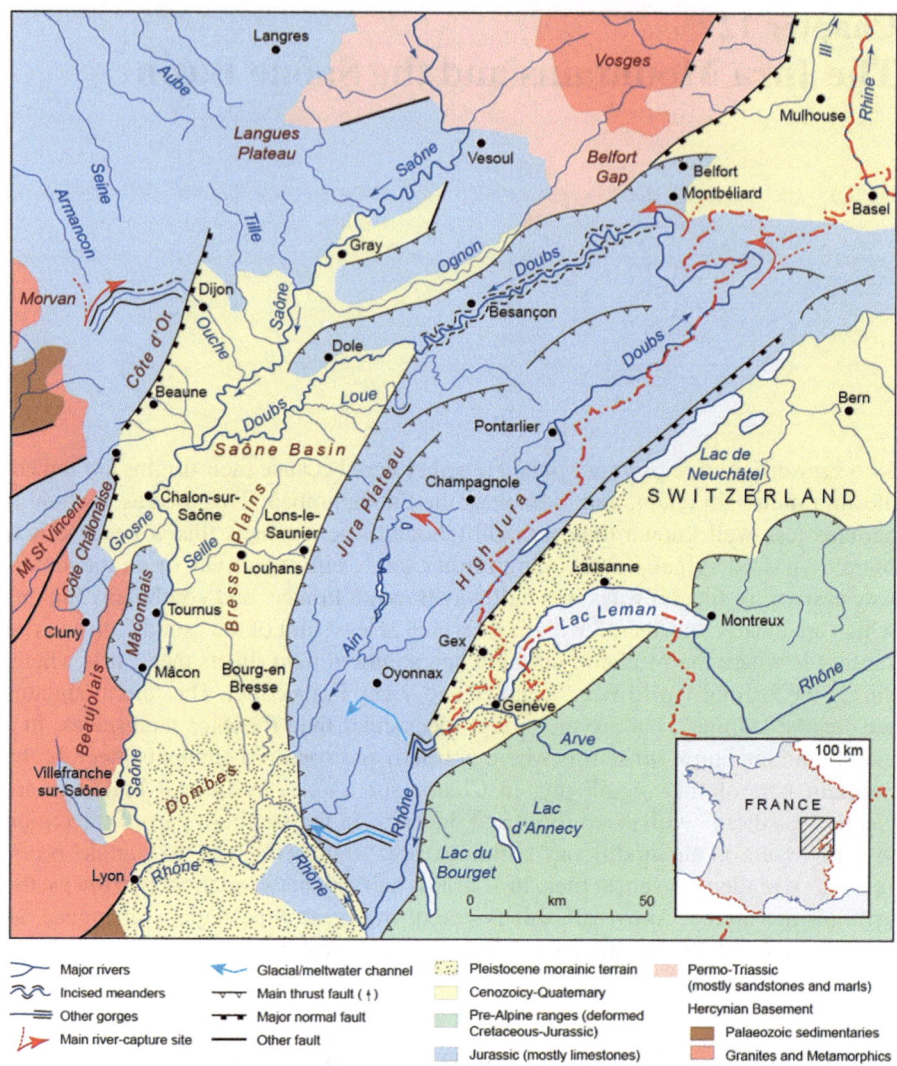

Fig. 12.1 Map of the Jura/Saône region (See also Figs. 3.7b Morainic terrain, The Dombes; 4.1c Limestone pavement, Jura; 4.1d Azé cave, Mâconnais; 4.3c Teracettes, Grosne valley; 4.5d Misfit meanders, Ognon valley; 4.8b Moraines, St Laurent, Jura; 5.1 Côte Chalonaise escarpment, Saules

explanation). Then, in the central unit, the Middle Jurassic limestones are little deformed and form a near horizontal but deeply dissected plateau. The western edge of this plateau is faulted, beyond which are a series of fairly tight and thrusted folds. The whole system has been thrust West towards and over the eastern margin of the Saône/Bresse graben, over a near-horizontal thrust plane developed in the

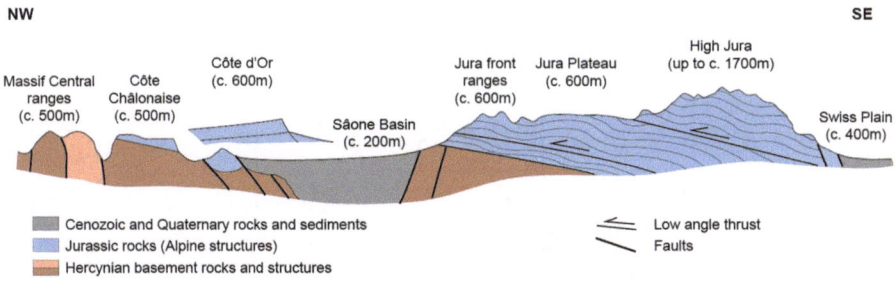

NW   SE

Fig. 12.2 Geological and topographic section across the Jura/Saône region

underlying saliferous Triassic marls. To the north of the Jura region proper (north of the town of Dôle, between the valleys of the Ognon/Saône and the Doubs) is a west-projecting promontory dominantly of folded Jurassic rocks which form the La Serre ridge (Fig. 12.1 map).

To the west of the Jura is the Saône/Bresse graben, a deep-seated tectonic depression, whose eastern margin is formed by the "Mid-Tertiary" (Alpine) thrust front of the Jura, and whose western margin marks the eastern boundary of the Massif Central (see Chap. 8). The graben floor is of Miocene sediments buried by Pliocene and Quaternary fluvial sediments and Pleistocene glacial and lacustrine sediments (Fig. 12.1 map). According to Buoncristiani and Campy (2011) during the penultimate (Riss) glaciation (c150 ka BP) the southeast corner of the graben (the Dombes area) was affected by a lobe of glacial ice derived from the middle Rhône valley, which created a very distinct morainic landscape (see below, see also Fig. 3.7b). Further north the ice limits were entirely within the Jura mountains (Fig. 12.3 map: see also below).

The northern limit of the Saône graben is essentially defined by the Middle Jurassic limestone of the Langres plateau. This limestone can be traced southwest where it forms the fault scarp of the Côte d'Or (above the classic Burgundy vineyards). The scarp can be traced south to form the Côte Chalonaise, where the limestone rests unconformably on Massif Central Hercynian rocks. East, across the Grosne valley from there are detached Jura-type structures in the Mâconnais ridges, standing above the Saône valley at Tournus. Are these ridges perhaps a southwestward continuation of the Jura structures present in La Serre ridge north of Dôle? The Mâconnais structures can be traced southwest from the Mâconnais itself to Solutré (see below and Fig. 12.6b), a spectacular monoclinal Jurassic limestone hill of major Mesolithic archaeological significance. Beyond Solutré the Jurassic rocks rest unconformably on the Hercynian Massif-Central granites of Beaujolais. Southwards from there the Beaujolais granites are faulted directly above the Cenozoic/ Tertiary rocks of the Saône graben.

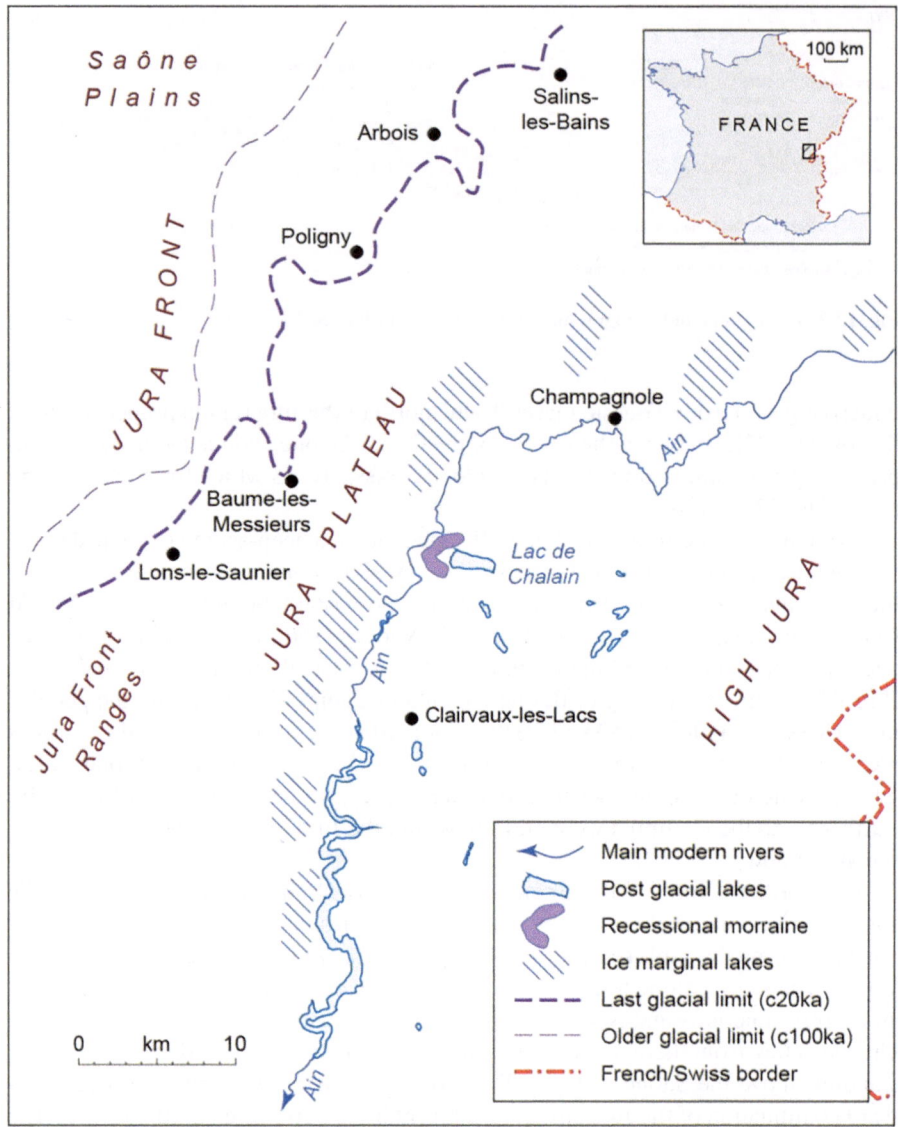

Fig. 12.3 Map of the Pleistocene glacial features within the Jura

The Geomorphology of the Jura/Saône Region

The Jura mountains form a series of northeast-southwest to north-south aligned ridges of folded Jurassic rocks. The highest ridge is to the east (maximum elevations just above 1700 m), standing wall-like above the western Swiss Geneva plain (Fig. 12.4a). North and east of there the ridges of the high Jura decrease in

Fig. 12.4 The Jura. (**a**) View across Lac Léman (Switzerland), from the flanks of the Jura (Col de la Faucille) towards the High Alps and the Mont Blanc Range. (**b**) A limestone gorge cut into the Jura Plateau, south of Arbois. (**c**) Late Pleistocene recessional moraines created small lake basins one of which is shown here: Les Quatre Lacs, central Jura

elevation but stand well above the western ridges (the so-called low Jura or Jura plateau Fig. 12.4b). This area in turn stands well above the Saône plains to the west and is deeply dissected by the Loue and Ain valleys. Northwards, the Jura plateau extends to the Belfort gap, separating the Jura from the southern flanks of the Vosges massif (see Chap. 7). Extending west from this northern part of the Jura plateau between the Ognon/Saône and Doubs valleys, is a promontory of Jurassic rocks and Jura structures, La Serre ridge, to the north of the town of Dôle (see above).

According to Buoncristiani and Campy (2011), the whole of the Jura mountain area, south of the Belfort gap, was glaciated during the maximum Pleistocene glaciation (the penultimate, the "Riss" glaciation, c150 ka BP) by a north-westward extension of the Alpine/Swiss ice cap (Fig. 12.3 map). At its maximum during that glaciation, the ice extended west to the foot of the Jura. During the last (Würm) glaciation (maximum c20 ka BP) the ice limit was within the Jura, creating morainic topography within the major valleys of the Jura (see Fig. 4.8b). That limit is marked by a string of former ice-marginal lakes extending from the upper Ain valley northeast towards Pontarlier (Fig. 12.3 map). Further to the east a series of small moraine-dammed lakes, including Lacs de Chalain, and lakes at Clairvaux-les-Lacs and Les Quatre Lacs (Fig. 12.4c) relate to a slightly later glacial recessional stage. Other evidence of Pleistocene glacial ice in this area lies southwest of the Jura in the southern part of the Saône graben (Bresse), the fluted ridged morainic country of the Dombes. According to Buoncristiani and Campy (2011) this morainic zone is of penultimate glacial (c150 ka BP) origin, formed by a lobe of ice emanating from the glacier that occupied the middle Rhône valley, in the Ambérieu area (see Chap. 13). The Dombes area is a series of low ridges, radiating in a fan-like form towards the northwest from the modern Ain/Rhône confluence zone. Between the ridges are a series of shallow lakes (see Fig. 3.7b), many now reclaimed, but a number still perennially wet, forming important ornithological sites.

As a limestone area the Jura is rich in karstic features, from small- to medium-scale solutional features (see Fig. 4.1c) to major underground drainage. Surface rivers are few and much of the drainage is underground, emerging at major springs such those as at the waterfall site at Hérisson. Other springs occur beyond the Jura margins, such as that at Bèze. Caves are also important and include "tourist" caves, such as those at Baumes-les-Messieurs. Within the Jura mountains themselves the only real rivers are the major rivers: the Ognon (Fig. 4.5d map) and the Doubs in the north, both with incised meanders, and the Ain, a north-south flowing river within the Jura also with incised meanders, but extensively dammed and now mostly a string of artificial lakes.

West of the Jura Mountains the floor of the Saône graben is relatively flat, dominated by the floodplain and low terraces of the Saône and its tributaries from the east, particularly the Doubs in the north and the Seille in the centre. Further downstream beyond Chalon-sur-Saône the river has a wide semi natural meandering channel within a wide floodplain (Fig. 12.5a, b).

The western edge of the Saône graben is defined by the series of faults that bound the eastern margin of the Massif Central (see Chap. 8). These are major regional faults, downthrown to the east and some having a throw of more than 300 m. The boundary comprises three distinct segments. In the north, in the Côte d'Or escarpment section, a series of NNE-SSW orientated faults has created a terrain of similarly aligned plateaux, ridges and valleys. The dominant resistant lithology is Middle Jurassic limestone. The scarp slope of the most easterly ridge (the Côte de Nuits, and further south the Côte de Beaune) supports some of the most famous Burgundy vineyards. At the mouths of small valleys that breach the scarp are Quaternary alluvial fans, adding variety to soils and therefore to the wine quality from the scarp-foot vineyards. Erosion along the series of faults behind and parallel with the main frontal fault, especially south of Beaune, has produced a series of valleys parallel to the frontal scarp. Behind (to the west) is the main scarp of the middle Jurassic Limestone (for example, above the spectacular château-site of La Rochepot). From the crest of the scarp, at a number of sites, there are spectacular views to the east, across the Côte d'Or limestone terrain and the Saône graben towards the Jura. It is said that on a clear day Mont Blanc can be seen from here too!

The middle segment of the western edge of the Saône graben, from Chagny to south of Mâcon, including the Côte Chalonaise and the Monts de Mâconnais, is more complex. The middle Jurassic limestone caps the fault scarp of the Côte Chalonaise south from Chagny to St-Gengoux-le-National. The limestone thins

Fig. 12.5 The Saône basin (**a**) The River Saône at Ouroux, a large regulated lowland river. Note the skyline to the left is the Côte Chalonaise Jurassic limestone escarpment. (**b**) Flood conditions on the River Saône, at Ouroux. (**c**) The River Grosne at Lalheue, a lowland meandering river, a right-bank tributary of the Saône, fed from the Cluny area and the Côte Chalonaise. (**d**) The River Grosne at Lalheue, in flood

southwards, so that the underlying Triassic sandstone and the Hercynian basement rocks are increasingly exposed at its base. South of Buxy the Côte Chalonaise (see Fig. 5.1) is fronted by the downfaulted small graben of the lower Grosne, filled with Tertiary and Quaternary clayey sediments. The Grosne itself is a small low-gradient meandering river (Fig. 12.5c, d). Eastwards across the Grosne lowland, are the Monts du Mâconnais, a series of faulted north-south ridges, composed of Hercynian basement rocks in the west, overlain by Triassic sandstones and a repeated series of Middle and Upper Jurassic limestones. This may be a southwestward extension of the frontal Jura structures seen on the other side of the Saône graben in La Serra ridge (see above).

Just south of Mâcon the Mâconnais structures are truncated by the fault-aligned valley of the Petit Grosne (followed by both the TGV line and the motorway-style main road from Cluny; Fig. 12.6a). South of the Petit Grosne valley are more Jurassic rocks, culminating in the dramatic Solutré Rock (Fig. 12.6b), a Middle Jurassic limestone-capped, eastward-dipping, scarp fragment. South of Solutré a fault truncates the Jurassic sequence placing the rocks directly against the Carboniferous volcanic rocks and intrusive granites of Beaujolais, part of the Massif Central. Beaujolais forms the third zone of the western boundary of the Saône graben (Fig. 12.6c). Its junction with the Saône graben is a fault, locally buried by Quaternary alluvial fans, which were derived mainly from the Beaujolais granites. The dominant land use is vineyard for the Beaujolais wines. Behind, to the west, the Monts du Beaujolais rise to almost 900 m, whereas the floor of the Saône graben

Fig. 12.6 West of the Saône. The terrain includes the Côte d'Or, Côte Chalonaise, Mâconnais, Beaujolais (see also Fig. 5.1, Saules; and the caves at Azé Fig. 4.1d). (**a**) View from the southern flanks of the Mâconnais ridges at Berzé-le-Château, looking south across the Petit Grosne valley into Beaujolais. The scarps in the centre distance are of the Jurassic limestones that surround Beaujolais. The Beaujolais granitic terrain is to the distant far right. (**b**) The Jurassic limestone scarp on the eastern edge of Beaujolais at Solutré: an important Neolithic archaeological site, as well as being an important vineyard. (**c**) In the heart of the dissected granitic terrain of the Beaujolais vineyards near Mt. Brouilly, to the west of Belleville

here is at c160 m. South of Beaujolais (near Villefranche) is another small area of Jurassic limestone terrain, but it rapidly gives way to the Hercynian metamorphic terrain which continues south into the Monts du Lyonnais (see Chap. 8).

The drainage of the whole area (both the Jura and the Saône graben) is dominated by the Saône system (Fig. 12.1 map). The River Saône itself rises in the southwest corner of the Vosges (see Chap. 7) then crosses Jurassic rocks within which incised meanders are well developed. Near Pontailler-sur-Saône, just within the Saône graben, it picks up an east-bank tributary, the Ognon (see Fig. 4.5d map), another river that rises in the southwest part of the Vosges, and whose middle reaches comprise incised meanders within Jurassic limestones, and which exhibit well-developed misfit modern meanders.

The major east-bank tributary of the Saône is the Doubs, an extraordinary river! (It is best appreciated on a map: Fig. 12.1 map). The Doubs rises in the southern Jura near the village of Mouthe, then flows northeast as a strike stream for about 8 km before a transverse section into the Lac de Saint-Point valley. It becomes transverse again to Pontarlier. From there it heads northeast for a while following the Swiss border, then it flows into Switzerland to St Ursanne where it turns abruptly to the west and back into France through St Hippolyte. There it turns north, crossing the structural grain to Montbeliard on the northern margin of the Jura. There it turns abruptly southwest through huge incised meanders through Baumes-les-Dames and Besançon, ultimately to Dole and out of the Jura into the Saône graben, joining the Saône at Verdun-sur-le-Doubs. The two big left-hand bends within the north-eastern part of the Jura are almost certainly sites of early captures (or glacial diversions?) by the Doubs system of originally Rhine drainage, but I do not know the details.

The Loue joins the Doubs just south of Dole, having drained the north central Jura from Ornans southwestwards. It has incised meanders in its upper course within the Jura. Further south the Seille, which joins the Saône near Tournus, is a sluggish stream, almost entirely fed by the eastern part of the Saône graben. The southwestern part of the Jura is not within the Saône drainage, but is drained by the north-south aligned Ain, a tributary of the middle Rhône. There are spectacular incised meanders in the middle reaches of the Ain, but upstream the river has been dammed to form the artificial Lac de Vouglans.

The Saône itself is a large lowland river with a dominantly meandering channel flowing through a wide modern floodplain. Transporting mostly fine suspended sediment, it is interesting to compare the Saône with the Rhône at their confluence in Lyon. The Rhône is an Alpine and sub Alpine-fed river, with higher velocities and a coarser sediment load (see Chap. 4). On the other hand, the Saône is a more sluggish river with lower velocities, supporting a finer (clay-rich) suspended sediment load. Although the Saône is managed for navigation with weirs and locks to allow the passage of moderate sized vessels (barges and river cruise vessels) at least as far upstream as Chalon-sur-Saône, it is less modified than other French rivers of a similar size.

The west-bank tributaries of the Saône include the Ouche, incised into the Jurassic limestone terrain, west of Dijon. At an early stage in its development the Ouche appears to have captured the headwaters of the Armançon (Seine drainage) in the area around Bligny-sur-Ouche (see Chap. 9). Further south the Dheune joins the Saône near Verdun-sur-le-Doubs. The Dheune drains the Le Creusot valley of the eastern Massif Central. South of Chalon-sur-Saône the next main tributary of the Saône is the Grosne (Fig. 12.5c, d), which drains the north side of the Beaujolais granite, before flowing through the Cluny valley and the lowland between the Côte Chalonaise and the Monts du Mâconnais. Within Beaujolais itself the streams are short and drain directly into the Saône south of Mâcon.

Highlights: The Jura

Col de la Faucille (Eastern Side) There is a Panoramic view across the Geneva lowland to the high Alps beyond, including the Mont Blanc Massif (Fig. 12.4a), on the D905 (the old N5 main road to Geneva) as the road descends from the high Jura into the Geneva lowland. This is a spectacular view over the western end of the central Swiss plain including the western end of Lac Léman and the Geneva lowland. It is backed by the Chablais Pre-Alps (of folded Jurassic rocks) behind which are the crystalline northern Alps including the Aiguilles ranges and above all a superb view of the Mont Blanc massif—at least when Mont Blanc is not cloud covered. I have been here on many occasions—on most of those occasions Mont Blanc was cloud covered. Localised cloud over Mont Blanc is not uncommon!

Karst Features of the Jura The Jura mountain range is one of the classic karst areas of France with a wealth of karstic features at all scales from small-medium scale solutional features (Fig. 4.1c) to large-scale limestone gorges (eg. the Baume-les-Messieurs valley in the Jura front ranges east of Lons-le-Saunier; and in the Ornans valley at the head of the Loue drainage). Included are karstic caves, several of which occur in the Ornans area. Other drainage features include sinks and resurgencies, waterfalls (eg. Hérisson waterfall), and karstic lakes (eg. Lac de l'Abbaye de Grandvaux).

Glacial Limits (Fig. 12.3 Map) The limit of the penultimate glaciation (c150 ka BP) is present in the Jura, and can be traced intermittently just back from the front of the Jura between Lons-le-Saunier and the Arbois area. Much clearer is the last glacial limit (c20 ka BP) that forms terminal morainic ridges along the upper valley of the Ain. Some of these ridges enclose lake basins, either at the maximum limit or a little to the east, marking glacial recessional stages, for example the Lacs de Chalain, Chambly, Clairvaux and Les Quatre Lacs (Fig. 12.4c). The Lac de Chalain also has archaeological significance as the site of a Bronze Age settlement of pile-supported dwellings.

Meandering Rivers Within the Jura, especially in the northern Jura, there are excellent examples of several types of meandering river, from incised meanders (including misfits) to partially confined incised meanders to open free meanders.

The Doubs: The Doubs in the Besançon area has excellent incised meanders, especially upstream of Besançon, confined meanders near Roulans, and a transverse reach at Clerval, but more open incised meanders near L'Isle-sur-le-Doubs. An interesting challenge, but one not directly related to incised meanders rather to long-term drainage evolution, would be to follow the extraordinary course of the upper Doubs from its source near the Swiss border downstream to Montbéliard, and to ponder on the possible causes for this extraordinary course. (Previous relationships with Rhine drainage but finally captured by Saône drainage?)

The Ognon (see Fig. 4.5d map): This river has amazing misfit modern meanders within older incised meanders, especially in the reach from Boulot to downstream of Marnay.

La Loue: A freely meandering lower reach (with preservation of active meander scrolls) near La Loye (southeast of Dole) contrasts with the confined transverse reach near Mouchard and the incised meanders in the uppermost reaches upstream of Chenecey-Buillon and near Ornans.

Highlights: Saône Basin

An Itinerary Through the Côte d'Or and the Classic Burgundy Wine District [A round trip from Dijon, including the "classic" wine route, also La Rochepot, the Côte d'Or escarpment and the aggressive Ouche drainage]. From Dijon take the D974 south to Nuits-Saint-Georges and Beaune. Throughout, the frontal of the two fault scarps of the middle Jurassic limestone is to your right with the classic wine districts on the talus and colluvial slopes below it. (A comment from one of our sons on his first visit to this area—"The road signs read like the labels on the shelves of an up-market off-licence/liquor store"!) From Beaune (which is worth a touristic visit in its own right) take the D973 through the first escarpment to La Rochepot (note: the château), then turn right onto the D906 to the top of the escarpment. (Note that the village of Gamay, after which a red-wine grape is named, is not far southeast of here). From the top of the scarp here (near Belair, Bout du Monde) there are panoramic views over the double escarpment itself, the Bresse plain and to the Jura beyond.

From the Bout du Monde continue north on the D906 for about 6 km to a right turn onto the D17 signed Cussy-la-Colonne (a Roman column), and Montceau-et-Écharnant, then continue north towards the Ouche valley. The Ouche, a tributary of the Saône, at some stage probably in the mid-Pleistocene, cut back to form an incised valley through the scarp to intercept and capture the low gradient north-flowing former headwaters of the Armançon, part of the Seine drainage (see Chap. 9). Beyond Montceau-et-Écharnant the road descends into the wooded incised valley of the Ouche to Lusigny-sur-Ouche, a charming village beside the spring-fed headwaters of the River Ouche. In Lusigny turn left onto the D970 to Bligny-sur-Ouche (with its tourist railway) then take the D33 down the incised Ouche valley. Initially the valley is fairly confined, but beyond the Château de Marigny it becomes less confined, exhibiting rather ill-developed incised meanders within which the modern River Ouche is a misfit. This valley eventually leads back to Dijon.

Côte Chalonaise—Route des vins Between Montagny and Culles-les-Roches From the delightful small town of Buxy, head west along the D977, and after about 1 km turn left into the village of Montagny-lès-Buxy which lies at the foot of the scarp of the middle Jurassic limestone of the Côte Challonaise. [Montagny is also the home of an excellent white Burgundy wine.] Several km to the west of Montagny within the faulted country on the margins of the Massif Central is Mont-St Vincent, a site which gives extensive views of that terrain (see Chap. 8, Highlights).

However, to continue with this excursion, from Montagny head south along "Route des Vins" at mid height along the escarpment. Just south of Montagny on the left of the road is a viewpoint with a "Table d'Orientation" from which there is an extensive view east to the Grosne valley, the north end of the Mâconais ridges and to the Saône plains and the Jura beyond. It is said that on about six occasions per year of exceptionally clear weather Mont Blanc is also visible. (I must have been lucky, I have seen it more than once!) From this site continue south along the "Route des Vins" through the villages of St Vallerin and Chenôves to Saules (see Fig. 5.1). There turn right to Culles-les-Roches, a village nestling immediately below the escarpment, and where there is an informative "geotrail".

This excursion can be completed by visiting St Gengoux-le-National, another delightful southern Burgundy small town. The simplest way is via the former railway station south of the village of Culles-les-Roches on a back road directly to St Gengoux. While in St Gengoux a visit to a hilltop viewpoint to the west of the town affords extensive views to the east over the Grosne valley and the Mâconnais ridges.

An Itinerary Through the Mâconais Ridges Including the Azé Caves The Mâconnais is an area of north-south aligned fault-bound ridges, composed of rocks ranging in age broadly from a base on Upper Palaeozoic through Triassic to Jurassic. Eocene rocks are also involved. The scarp formers are mostly Middle Jurassic limestones.

From Mâcon head west from the town along the D17 to La Roche-Vineuse. There, turn north along the D85 which runs along one of the S-N fault-guided valleys through Verzé and Igé to Azé. The valley-side slopes form part of the Mâconnais wine area. The Azé caves are certainly worth a visit (see Fig. 4.1d). They have archaeological, palaeoenvironmental as well as karstic geomorphological interest. They are becoming an ever more popular tourist attraction, so especially in peak season you need to book in advance.

From Azé drive north on the D82 then the D161 to and through Cruzille. The road follows the fault-aligned valley, whose side slopes are mantled by Mâconnais vineyards. (Note that the village of Chardonnay, after which the white wine grape is named, is a little off to your right.) Beyond Cruzille continue on the D161 to Brancion, a restored medieval village worth a visit for its own historical interest, but there are also scarptop views to the west over the Grosne valley towards the fault scarp of the Côte Chalonaise in Jurassic limestone. From Brancion, turn east onto the D14 through Ozenay, crossing several low faulted N-S ridges in Jurassic limestones to reach Tournus, on the banks of the Saône. Tournus itself is well worth a visit.

Southwest of Mâcon, the Solutré Area (Fig. 12.6a, b) This small area to the southwest of Mâcon has two claims to fame. First it is the home of Puilly-Fuissé white Burgundy wine. Second it is an archaeological site of international importance. In addition, the landscape is impressive.

Two tabular escarpments of steeply northeasterly dipping Middle Jurassic limestones form the Vergisson and Solutré rocks. The Solutré rock (Fig. 12.6b) has the greater archaeological significance. The summit comprises a wedge-shaped limestone rock outcrop, below which a gentler slope is punctuated by smaller rock ledges. At the base of the slope are lower-angle segments comprising periglacial materials. It is within these materials that significant archaeological remains were found, particularly numerous horse bones together with numerous flint tools. The name Solutréan was given to this Palaeolithic culture, which was subsequently dated to about 20–17 ka BP. The site though has been a site of human activity both prior to that and since. There is a superb archaeological museum at the foot of the scarp—certainly worth a visit. To locate the site leave Mâcon heading south on the D906, but once out of town turn onto the N79 expressway towards Cluny. After a few km turn off following the signs for Pouilly and Solutré.

Beaujolais (Fig. 12.6c) This is the famous wine growing district based on Hercynian rocks, mostly Carboniferous granite, but to the west are Carboniferous volcanic rocks and schists. The forested summits on the western watershed range in elevation between 700 and 900 m. The Saône floodplain ranges between about 170 m and 150 m between Mâcon and Villefranche, less than 20 km to the east. So the Beaujolais terrain is relatively steep, dissected by a dendritic network of small streams. Below the forested western summits most of the area is devoted to vine cultivation. I would suggest a route starting from Belleville on the Saône, heading northwest on the D18 through Villié-Morgon (famous for Beaujolais Morgon wines) towards Avenas. The road climbs off the granite onto the metamorphic and volcanic rocks, leaving the vines behind and on into forest. As you climb through hairpins onto a plateau (at the Col du Truges) there is a magnificent viewpoint with a "Table d'Orientation" facing east across the Beaujolais vineyards towards the Saône valley and the Bresse plain towards the Jura, and, with reasonable visibility, to the Alps beyond including the Mont Blanc range.

Return down the hairpins back into the vineyards to the crossroads at Le Truges. Here turn very sharp right onto the D26 towards Beaujeu. At Beaujeu you actually want to turn left onto the D37 that follows the valley of the Ardière (a small meandering stream) southeast from Beaujeu towards Mont Brouilly. You could take a short cut from the D26 to the left through Régnié-Durette onto the D37, but be warned—Beaujolais is riddled with lanes and minor roads—it is very easy to get lost! However, follow the D37 southeast to where it intersects with the D43. There, turn right (south) onto the D43 (signed for Villefranche). After about 1 km there is a minor road on the left to Mont Brouilly, another excellent viewpoint. From the viewpoint return to the D43, turn left (south) towards Villefranche.

The Dombes (see Fig. 3.7b) At the peak of the penultimate glaciation, according to Buoncristiani and Campy (2011), a huge glacier spilled west towards Lyon, fed from the northern French Alps, the Geneva area, and the southern Jura. Beyond the southern Jura it divided into two lobes, one flowing down the Rhône axis towards Lyon itself (reaching to about Montluel, east of Lyon). The other lobe spilled northwest over the low watershed at Meximieux into the southern part of the Bresse plain, radiating out in a great arc north, northwest and west, and depositing glacial sediment over that area. As the ice melted it generated large amounts of meltwater, almost certainly sub-glacially. The ice and meltwater grouved the sediment into shallow valleys radiating out in fan-like form from an apex near Meximieux. On deglaciation, hollows within these valleys became shallow linear lakes. Many of these have now dried out or have been drained for agriculture. Others have become ephemeral shallow lakes in winter, marsh in summer, but there are still others that are open water. An important present-day function of the Dombes is as a haven for water birds, so if ornithology interests you a visit could be worthwhile—either informally or specifically to the ornithology park at Villars-les-Dombes. To see something of the Dombes landscapes, the area within several km north and south of the Villars to Chalamont D904 road would be suitable.

The Upper Saône—A Large Meandering River, Plus Misfits In comparison with most other large French rivers, the Saône is relatively little directly modified by engineering. Downstream from Chalon however, there are locations with bank protection, a few artificial cutoff channels and the flows are regulated by wiers and navigation locks. Upstream from Chalon the picture is similar, but the extent of human intervention is less. Several reaches are interesting.

From Port-sur-Saône downstream to Gray: This is the most upstream reach within which the Saône could be considered a major river. For much of this reach there are large valley meanders within which the modern channel is underfit. There are a few navigational by-pass channels, otherwise the river has a mostly natural channel. Former abandoned palaeochannels are evident in the floodplain.

From Pontailler-sur-Saône to Auxonne: This is a moderately natural meandering channel with only one local navigational cutoff.

From St Jean-de-Losne to Suerre: There is a navigational channel by-passing tortuous meanders (between Pagny and Suerre) Several natural cutoffs are also evident.

Chapter 13
The Northern French Alps and the Middle Rhône

This Chapter deals with the northern French Alps, east to the Swiss and Italian borders, north along the middle Rhône valley (ie. south of the Jura, see Chap. 12 above), west to the edge of the southern part of the Rhône/Saône graben, and south to the Drôme valley, the Écrins massif and the uppermost part of the Durance valley (Fig. 13.1 map). I deal with the Alpine area south of that in the next Chapter (Chap. 14). There are several reasons for this treatment of the Alpine area. First, the geology differs. All but one of the major crystalline massifs (the Mercantour massif) are in the northern sector. (Both the small Queyras Massif and particularly the even smaller Ubaye Massif are really on the border between North and South, Fig. 3.2. map) The Mesozoioc stratigraphy of the Pre-Alpine ranges also differs between north and south. In the north the Lower Cretaceous (Urgonian) reef limestone has a major influence on the morphology. It is much less important in the southern Pre-Alps. On the other hand the Upper Jurassic black shales that erode into the gullied and badland terrain of the "Terres Noires" of northern Provence, are of major importance in the south. The structures also differ: in the north they are dominated by "Alpine" mostly arcuate north- south structures. In the south such structures interact with east-west "Pyrenean" structures (Fig. 3.2 map). The Quaternary and modern characteristics also differ. The Pleistocene Alpine ice cap was virtually restricted to the northern area (Fig. 3.6 map), but in the south the only glacial ice, apart from that in Mercantour, was a penultimate glacial was an outlet glacier tongue extending down the Durance valley to somewhere short of Sisteron. Modern glaciers are restricted to the northern crystalline massifs. The present climate also differs, with the northern area having an "alpine continental" climate, but that of the south having a distinctive Mediterranean influence. The drainage too differs, the north dominated by Rhône and Isère drainages (Fig. 13.2 map), the south by the Durance system (Fig. 3.4 map).

A. Harvey, *The Geomorphology of French Landscapes*,
https://doi.org/10.1007/978-3-031-68490-6_13

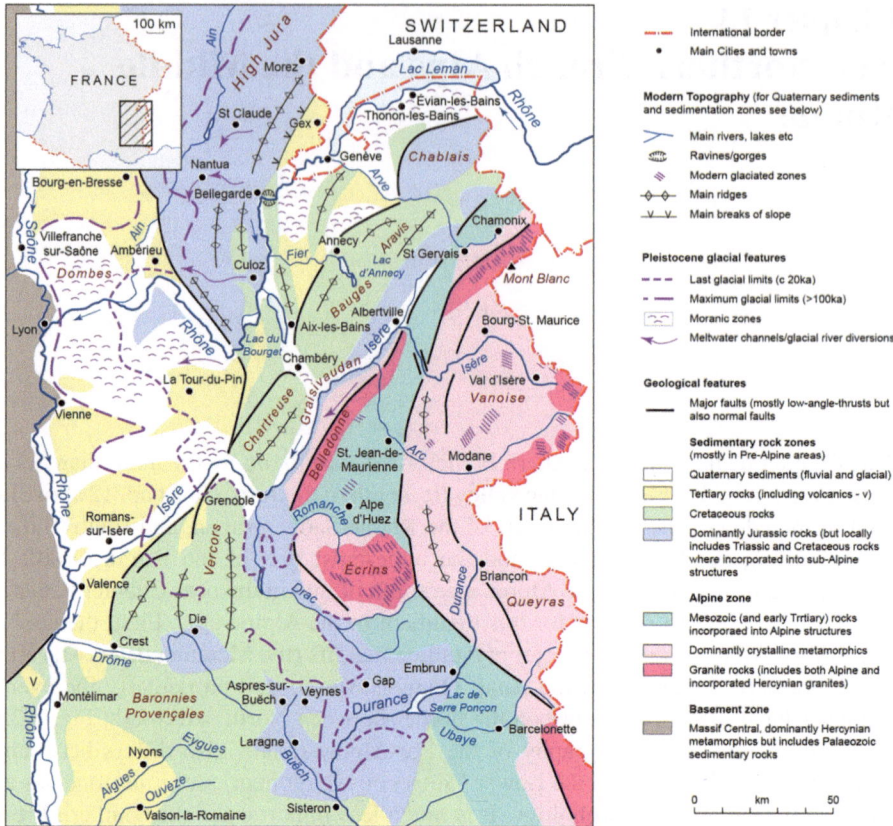

Fig. 13.1   Map of the northern French Alps and the middle Rhône valley

Geology of the Northern French Alps

The core areas of the northern French Alps comprise a series of belts of mostly metamorphic rocks (including bodies of Hercynian rocks and structures eg. the Mont Blanc massif), plus intruded granites. The whole set is thrust by a series of (arcuate) thrusts towards the northwest (Fig. 13.1 map). These are the (high) mostly crystalline Alps, forming several discrete ranges, from North to South: the Mont Blanc, Belledonne, Vanoise, Grandes Rousses, Écrins, Queyras, Ubaye and Mercantour massifs, though the last is actually within the southern part of the Alps (see next Chapter). The structures themselves are incredibly complex, involving overfolded nappe structures thrust westwards over near-horizontal thrust planes, later truncated by sub-vertical fault planes.

To the west of these ranges (also thrust towards the west) are the more or less discrete Pre-Alpine ranges. The most easterly, forming the Chablais range on the Swiss border, south of Lac Léman, has affinities with the Italian Pennine Alps and

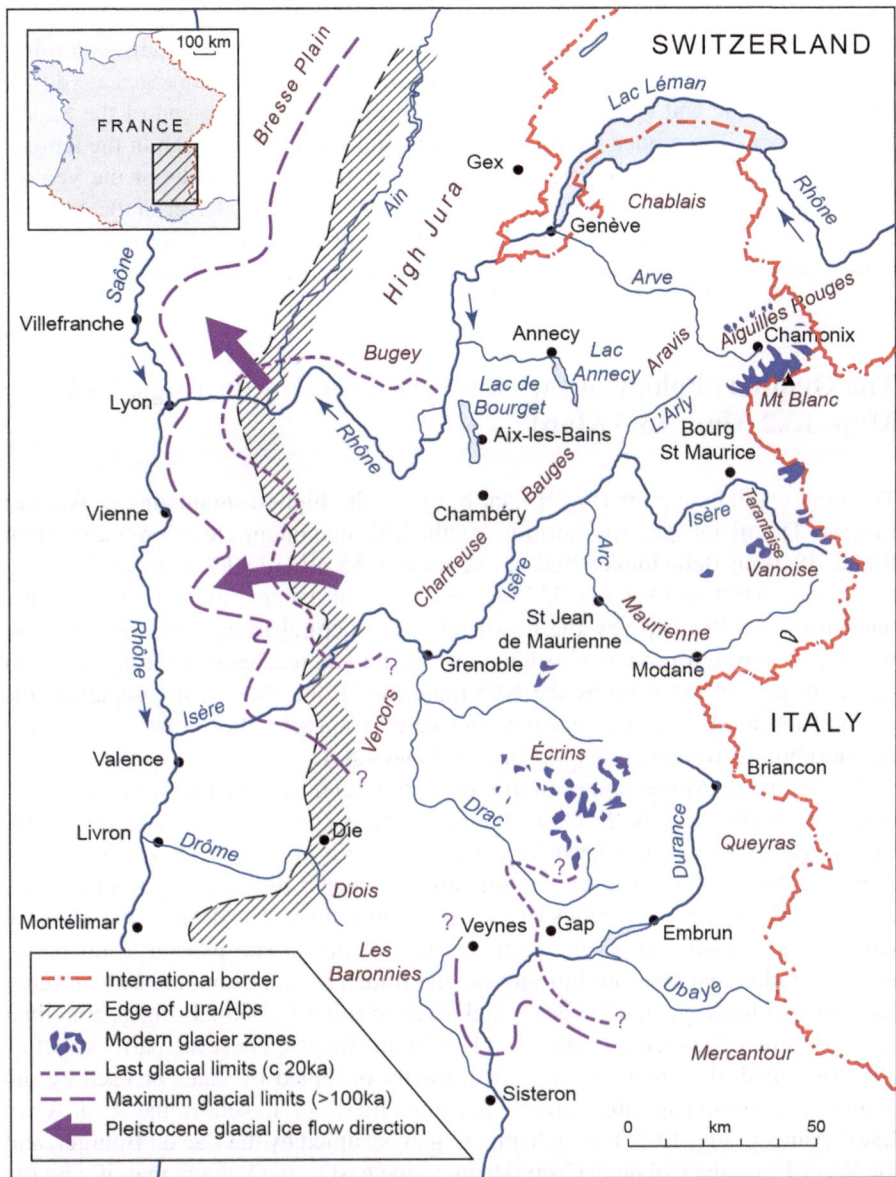

Fig. 13.2 Map of the modern river systems and the Pleistocene glacial limits in the northern French Alps

the Bernese Oberland in Switzerland. Otherwise the Pre-Alps in the northern Alpine section comprise sedimentary massifs dominantly of Cretaceous limestones, especially the Lower Cretaceous Urgonian reef limestone. These massifs occur as discrete structural and topographic units in an arcuate belt around the high crystalline massifs (from north to south: the Bornes-Aravis, Bauges, Chartreuse, and Vercors Massifs). They are separated from the crystalline massifs by a structurally-controlled linear depression, the Grésivaudin. To the west of the Pre-Alpine ranges are down-faulted Cenozoic and Quaternary sediments filling the southern end of the Saône/Rhône graben. The graben itself pinches out southwards (Fig. 13.1), in the latitude of the Drôme River, by the encroachment of the Cretaceous rocks of the Vercors Massif from the east with their continuation across the Rhône River in the Ardèche Plateau (see next Chapter). The western boundary of the Saône/Rhône graben is the faulted edge of the Hercynian rocks and structures of the Massif Central.

The Geomorphology of the Northern French Alps (Figs. 13.1 Map, 13.2 Map, 13.3 Map)

The high crystalline Alps (Fig. 13.4a, b, c) are the highest mountains in Western Europe. The major massifs (north to south, with maximum elevations) are: Mont Blanc, 4810 m; Belledonne, 2928 m; Vanoise 3855 m; Grandes Rousses, 3327 m; Les Écrins, 4102 m; Queyras, 3325 m. For much of the area the massifs are separated from the Pre-Alps by the thrust-related structural trough, the Grésivaudin, which is followed for much of its length by the middle reaches of the Isère River. In the south the col followed by the N85 road, the "Route Napoléon", separates the Écrins from the Pre-Alpine Dévoluy and Vercors massifs. (For other views of the geomorphology of these areas see Figs. 1.2 and 4.8a).

The main Pre-Alpine ranges (north to south, with maximum elevations Fig. 13.1 map) are: Bornes-Aravis, 2752 m); Bauges, 2158 m (Fig. 13.5a see also Fig. 4.3b); Chartreuse, 2062 m (Fig. 13.5b); Vercors, 2341 m (Fig. 13.5c, d); Diois, 1570 m and Dévoluy, 2587 m. All of these massifs are dominated by Cretaceous limestones, especially the Lower Cretaceous Urgonian reef limestone. The detailed morphology within these massifs ranges between simple escarpment and plateau forms dominated by individual resistant limestones, and folded terrain of both resequent relief (anticlinal ridges, synclinal valleys) and inverted relief (synclinal ridges, anticlinal valleys). Major gaps separate the main Pre-Alpine ranges. These are partly structurally controlled, then glacially modified, finally occupied by lakes or used by the main rivers. These gaps are currently occupied by (north to south): the River Arve; Lac d'Annecy (Fig. 13.5a); the Chambery gap occupied by the Lac du Bourget; and the River Isère; the Col de la Croix-Haute to the east of the Vercors massif; and the River Drôme, which flows in a great bend between the Vercors and Diois massifs.

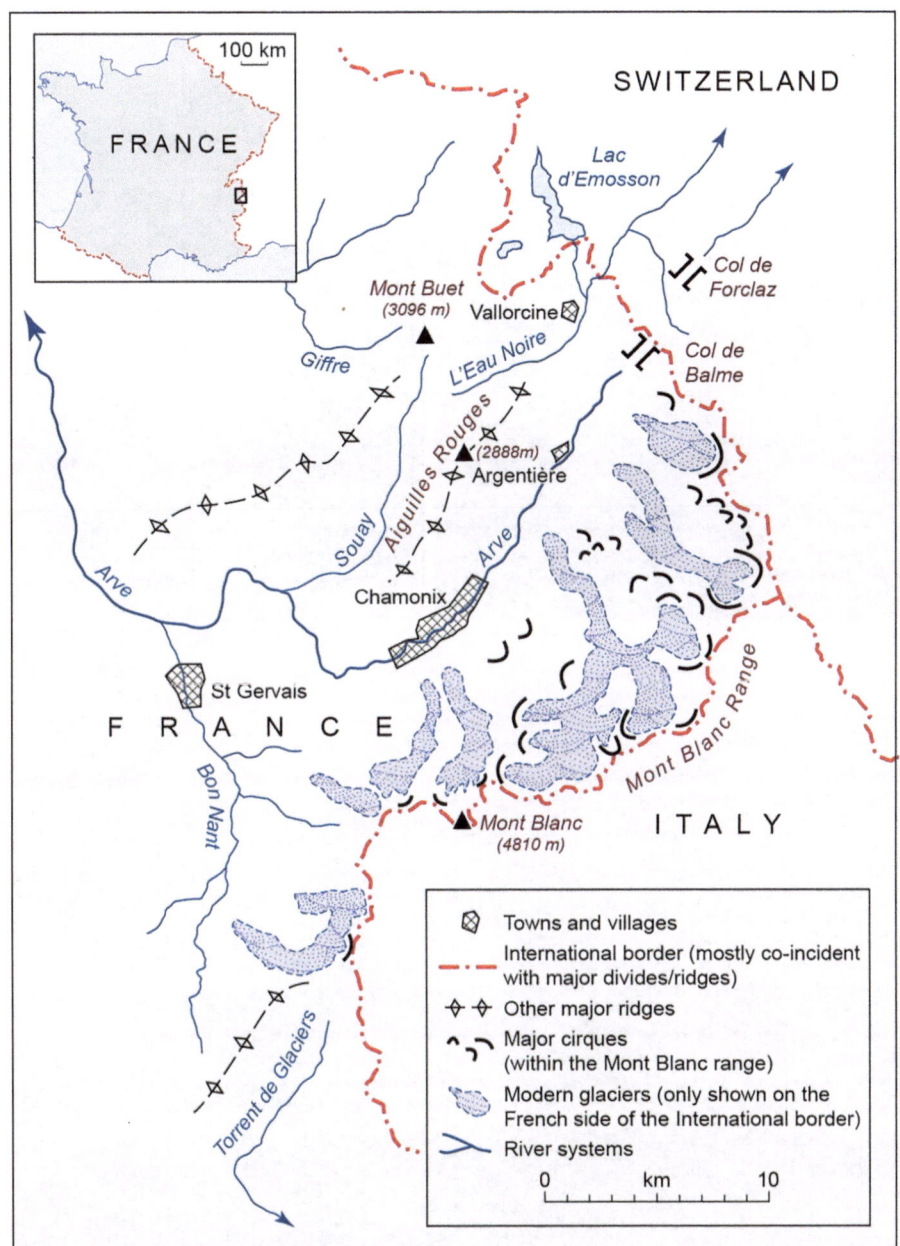

Fig. 13.3 Map of the modern glaciers in the Mont Blanc range

Fig. 13.4 The northern French Alps. (See also Figs. 1.2a Mont Blanc; 1.3a Argentière glacier, Mont Blanc; 4.4c Les Écrins, waterfall; 4.6a Col des Aravis, alluvial fan; 4.8a Mer de Glace). (**a**) Les Écrins from the north, from near La Grave. (**b**) The steep north face of Mont Blanc: Seen from across the Chamonix/Arve valley on the Merlet Road. Note the small perched glaciers. (**c**) The weather is not always perfect for viewing the north-face glaciers from the Merlet road!

Fig. 13.5 The northern Pre-Alps (including the Vercors Massif). Low mountains mostly of Mesozoic (especially Cretaceous) sedimentary rocks (see also Fig. 4.3b Sambuy Mountain, debris flows). (**a**) Lake Annecy: Occupying a glacially scoured double basin in the folded Cretaceous limestones of the northern part of the Bauges Massif. (**b**) The massive Cretaceous limestone of Mount Grenier, Chartreuse Massif, from which an enormous catastrophic rock failure occurred in 1248 AD. (**c**) The upper Drôme valley (in Diois) near the Claps landslide within the Vercors/Diois reach: Here the Drôme is a single-thread gravel-bed river, cut into Jurassic marls. This view is looking west (downstream) towards the reach in massive Lower Cretaceous limestone (beyond). The landslide debris (see text) is in the middle ground. (**d**) The lower Drôme valley near Saillans: Downstream from the Vercors reach the river has a wide channel with a tendency to braid

Glacial Geomorphology of the Northern French Alps

During maximum Pleistocene glaciation (the Riss Glacial stage at about 150 ka BP) a regional ice cap was centred on the northern Alps, extending west to the Dombes area north of Lyon (see Chap. 12, see also Fig. 3.7b) and to Lyon itself (Fig. 13.2 map, see also Fig. 3.6 map). It extended down the Isère valley to the western edge of the Rhône valley. In the south a glacier tongue extended into the southern Alpine area perhaps just to Sisteron (see Fig. 3.6 map). During the last glacial maximum (the Würm, about 25–20 ka BP) the ice cap was less extensive reaching down the Rhône only to Ambérieu, the Isère only to Voirons, and the Durance ice tongue only to somewhere north of Sisteron (see Fig. 3.6 map). During glaciation glacial erosion within the mountains would have been intense, scouring major troughs along the main ice streams. The main ice streams transported sediment to the glacial margins, dumping it as morainic areas (eg. the Dombes area in the corner of the Saône graben, see Chap. 12; and Fig. 3.7b). Glacial sediment was also dumped on the extensive plain between the modern Rhône and Isère valleys southeast of Lyon (Fig. 13.2 map). Beneath and beyond the ice, meltwater would have been an effective agent for erosion and deposition. Meltwater was particularly important within the Rhône/Isère plain, creating a series of east-west palaeochannels across the plain. Some of these may have been initially glacially scoured during the penultimate glaciation. Others may have been palaeo-Isère channels. Beyond the ice limits and affecting any nunataks above the level of the ice, permafrost would have been intense with its associated periglacial processes (see Chaps. 3 and 4).

As the ice cap shrank to a system of valley glaciers, erosion may have been even more concentrated. It created deep basins some of which on deglaciation became lake basins. In addition it created hanging valleys where the main glacial valley was deepened below the level of tributary valleys (see Fig. 4.4c). Deposition would have been focussed around the contemporaneous ice limits, creating cross-valley recessional moraines. At the same time discrete cirque glaciers would have fed ice to valley glaciers. Further reduction in ice volume would have increased the relative importance of small cirque glaciers and eventually resulted in complete deglaciation of most previously glaciated areas.

Modern glaciation is restricted to the highest mountain ranges (see Fig. 3.6 map) especially within the high crystalline massifs. Most of the modern glaciers are within the Mont Blanc range (Figs. 13.3 map, 13.4b, c: see also Figs. 1.3a, 4.8a), including both cirque and valley glaciers, but there are other small glaciers in the Vanoise and Écrins ranges (Fig. 13.4a) and a few glaciers in Queyras. If you are interested in modern glaciers then the Mont Blanc massif is a must. At a minimum, for a superb overview of the valley glaciers on the north face of Mont Blanc, go to the parking area of the "Parc Animalier de Merlet" on the opposite valley side (Fig. 13.4b, c)(see "Highlights" section below). For a closer view of a glacier, visit the Mer de Glace, beyond Chamonix (see Fig. 4.8a; See also "Highlights" section below).

Other Aspects of Modern Geomorphic Systems in the Mountains

Within the high mountain environments modern geomorphic processes continue to be highly active. At high altitudes, such as on the mountain ridges, mechanical weathering by frost action was an important process during the Late Pleistocene, and continues as such today. Sediment generated at high elevations moves downslope forming screes. Some of this material reaches small steep mountain drainages. Other sources of sediment may be hillslope debris flows (see Fig. 4.3b) feeding sediment downslope onto debris cones or into the stream system. That sediment, together with any generated within the headwater streams themselves, may accumulate where the underlying gradient decreases, forming alluvial fans (see Fig. 4.6a). Such depositional sites are common on the margins of formerly glaciated troughs.

Steep stream channels, especially where they are cut into bedrock (Fig. 13.6a) may be another source of fresh sediment. Further downsystem the mountain stream channels tend to be wide, shallow, gravel-bed channels (Fig. 13.5d), often with braided channel patterns. The geomorphology of the main river systems of the northern Alps is dealt with below.

Modern Drainage

The modern rivers draining the northern Alps feed either into the Rhône or the Isère (Fig. 13.2 map). The Arve, which drains the Chamonix/Mont Blanc area, carries a large glacially derived sediment load and joins the Rhône in the city of Genève (Switzerland), just below the Rhône's exit from Lac Léman. Contrast the sediment-rich waters of the Arve with the clean waters of the Rhône itself at that point. The fluvial geomorphology of the Arve, like that of many Alpine rivers is disappointing.

The channel of the Arve has been modified—straightened and channelized throughout the French reach. In the short Swiss there are meanders set below late Pleistocene terraces. Throughout its Swiss section the Rhône also meanders below late Pleistocene terraces. Once in France the Rhône passes through gorge reaches

Fig. 13.6 The northern Alps: rivers. (See also Fig. 4.4c Les Écrins, waterfall). (**a**) A boulder and bedrock channel within the Guil Canyon in the Queyras Massif. (**b**) The Jura section of the River Rhône at Lagnieu: A single-thread channel, looking upstream. The Jura form the skyline

through the southern Jura as far downstream as Seyssel, but from there down to and beyond Bellay, it is totally manipulated for hydro-electric generation. Beyond the big valley loop at the southern tip of the Jura, although the main channel has been duplicated, it is still possible to trace abandoned meander loops through what once was an amazing anastomosing reach. Former late Pleistocene meltwater channels can be traced from this reach across the plain towards Lyon. The modern river though heads northwest through a confined reach through Lagnieu (Fig. 13.6b) then west in what is mostly a single thread channel with occasional islands, to Charvieu-Chavagneux. Here the Rhône is joined by the meandering channel of the Ain that drains the southern corner of the Jura (see Chap. 12). From this confluence, downstream to Lyon, the Rhône is manifestly modified for hydro-electric generation. Within Lyon, at the south end of the city, the Rhône is joined by the Saône. Again, it is interesting to note the contrast between the slow-flowing, sediment-poor Saône (basically a lowland river, with only mud dominated suspensed sediment: see Chap. 12), with the faster flowing Rhône, even here still charged with coarser suspended sediment.

From Givors, south of Lyon southwards to Condrieu, the Rhône has a single thread channel within an incised valley on the margins of the Massif Central. Southwards from Condrieu the valley widens but again the channel has been radically altered for hydro-electric development. Similarly from Andance to Tain-l'Hermitage/Tournon (Côtes du Rhône country) the river is in an incised valley on the margins of the Massif Central, but the channel has been modified. Likewise from the Isère confluence, through Valence to the Drôme confluence at Livron, although this reach of the valley is less confined, most of the channel is radically altered.

From Lyon south to the Drôme confluence the Rhône picks up right-bank tributaries draining the eastern flanks of the Massif Central: the Gier at Givors; the Cance and the Ay south of Andance, both small streams with incised meanders; the Doux near Tournon, with its gorge and incised meanders; and finally opposite Livron, the Eyrieux and the Ouvèze, both within incised valleys. The left-bank tributaries of the Rhône that drain the low partially morainic country southeast of Lyon, are mostly small, most with sinuous to meandering channels. Only the Varèze and the Galaure are of any size.

Most of the northern Alps form part of the Isère drainage basin. The Isère itself is sourced above Val d'Isère between the Pennine Alps on the Italian border and the Vanoise massif. It then flows through the Tarantaise valley around the north side of the Vanoise massif through Bourg-Saint-Maurice to Moûtiers. From there it turns north to the head of the Grésivaudin at Albertville. A few kilometres along the Grésivaudin at Chamousset, the Isère is joined by the Arc from the Maurienne valley. The Arc is the major tributary of the Isère. It rises just over the Col d'Iseran from the headwaters of the Isère at Val d'Isère, then flows through the Maurienne valley around the south of the Vanoise massif westwards through Modane and St Jean-de-Maurienne then northwards to Aiguebelle to join the Isère at Chamousset in the Grésivaudin. Beyond there the Isère continues to follow the Grésivaudin past Pontcharra and the Chambéry gap to the north (between the Pre-Alpine Bauges and

Chartreuse massifs). Also at that point, across the valley on the edge of Chartreuse, is Mont Granier. That mountain is capped by Lower Cretaceous Urgonian limestone (Fig. 13.5b), on which there was a major rockfall and landslide in 1248 AD, the debris from which part buried the northeastern flank of the Chartreuse Massif.

The Grésivaudin continues south to Grenoble, where the Isère is joined from the south by its other main tributary, the Drac. The Drac/Romanche system, via the Romanche, drains the crystalline Galibier/Grandes Rousses ranges to the north of the Écrins massif. The south of Les Écrins and the zone between there and the Pre-Alpine Dévoluy and Vercors massifs is drained by the Drac. Beyond Grenoble the Isère runs northwest through the gap between the Chartreuse and Vercours massifs to Moirans, where it turns southwest to follow the edge of the Vercors massif almost to Romans, where it turns west to its confluence with the Rhône, a few km north of Valence.

From a geomorphological point of view, the fluvial landforms within much of the Isère system are disappointing. It is true that the headwater reaches of both the Isère and the Arc are natural and spectacular with a wealth of currently active landforms within recently (Late Pleistocene) deglaciated landscapes. The Isère valley above Val d'Isère includes active debris cones and small tributary-junction alluvial fans. Below there the Isère channel is an active braided channel. Similarly, over the Col d'Iseran, the upper Arc valley above Lanslebourg-Mont-Cenis is a recently deglaciated natural Alpine valley. The valley floor is flanked by recent stream terraces and active tributary-junction fans below which is a natural gravel-bed stream channel, braided in its upper reaches. Then, below Val d'Isère, through Tignes, the River Isère is dammed to form the Lac du Chevril. Then through Bourg-Saint-Maurice to Moûtiers, the river intermittently has a bedrock channel within shallow gorge reaches alternating with local valley widenings. In these wider reaches there are intermittent river terraces and local tributary-junction fans. From Moûtiers downstream to the Grésivaudin at Albertville, the channel is mostly constrained artificially, though there are a few semi-natural braided reaches, such as at La Bâthie.

The upper River Arc, from Termignon downstream to Bramans, has a modern floodplain with a sinuous gravel-bed channel, below recent low terraces, with local tributary-junction alluvial fans. Beyond there it is mostly a bedrock channel as far as Modane, then throughout the middle and lower Maurienne valley much of the channel has been severely modified by constraining walls, weirs and dams. There are a few semi natural reaches (eg. a tributary-junction fan and local braids upstream of St Julien-de-Maurienne). From there downstream to the confluence with the Isère the channel is highly modified with constraining walls, weirs etc.

Throughout the Grésivaudin from Albertville virtually to Grenoble the channel of the Isère is artificially straightened and constrained by floodwalls. At Grenoble it is joined by the Drax/Romanche system. Both of these rivers have also been radically modified. For example, the Drax has been dammed in numerous places and is now almost wholly a string of artificial lakes.

For several kilometres downstream from Grenoble the Isère has an artificial channel, beyond which, it appears to have a relatively wide, partially constrained channel, within which there are gravel bars. There is evidence in the floodplain of high sinuosity palaeochannels. Further downstream are several hydro-electric weirs or dams, for example at Saint-Nazaire-en-Royans there is a dam backing up the flow in what appears to have been a narrow bedrock controlled reach. Across the plain through Romans-sur-Isère to the Rhône confluence are several other dams/weirs. Also within the floodplain there traces of sinuous palaeochannels.

The Drôme, the most southerly of the "northern" Alpine rivers, drains the Pre-Alpine area between the Vercors and Diois, both massifs dominantly of Cretaceous Urgonian limestone (see above). The fluvial geomorphology of the River Drôme itself shows some remarkable contrasts between the upper reaches upstream of the great bend through Die, and the lower reaches downstream of Saillans. The upper reaches are of a dominantly single-thread channel, locally undermining the base of the valley-side slopes. At one site, Le Claps, south of Luc-en-Diois, the river cuts into bedrock at the base of a transverse shale and limestone ridge. This has triggered a major slope failure/landslide, whose debris forms a zone of "chaos" on the valley floor (Fig. 13.5c). From the Claps landslide through Die to Saillons, through the reach of the "great bend", the channel is mostly a sinuous single-thread channel that locally braids. There are local transverse bedrock ridges, especially upstream of Saillans. Downstream from Saillans (Fig. 13.5d) the river has a clear tendency to braid, a characteristic that is very pronounced in the final reaches across the plain between Crest and the Rhône confluence beyond Livron.

A final word in relation to visiting the northern Alpine region: there is a lot of spectacular, impressive and interesting geomorphology—far more than I can cover in detail here. To see it best try and avoid the main roads on the valley floors—there is a lot of traffic. It is often difficult to park or pull off, and you see much less than you can from higher up. The cols and passes on the minor roads (eg. the Col d'Iséran, see above, but also try the Cols de la Croix de Fer, Glandon and Galibier south of and above Saint-Jean- de-Maurienne—see Highlights below), offer much more impressive views, usually of landscapes much less modified by human activity than those seen from the main valleys.

Highlights of the Northern French Alps

Chamonix/Mont Blanc (see also Deline and Ravano, Chap. 17 in Fort and André, Eds. 2014) For geomorphologists, especially those interested in modern glaciers the Chamonix area is a must. Not only is Mont Blanc the highest mountain in Western Europe, but it is also home to the greatest concentration of modern glaciers (see above). They occupy both the northern (French) flank of the mountain mass and the southern (Italian) flank.

A whole series of short and steep valley glaciers occupy the north flank of Mont Blanc (Fig. 13.3 map), including the Taconnaz and Bossons Glaciers above Chamonix, the Mer de Glace (see Fig. 4.8a) to the east of Chamonix and the Glacier d'Argentière above Argentière further east (see Fig. 1.3a). An excellent overview of the series can be obtained from the parking area for the small zoo on the north slope of the Chamonix valley, the Parc-Animalier de Merlet (Fig. 13.4b). Choose better weather than on my last visit there (Fig. 13.4c). For a closer view of a glacier, the Mer de Glace (see Fig. 4.8a) is directly accessible. From the viewpoint above the glacier, note the details of the crevasse patterns within the ice, the dirt bands and the marginal moraines. The glacier has shrunk considerably over the last 10 years.

Lac d'Annecy (Fig. 13.5a) A beautiful Alpine lake, set within a glacial trough last occupied by a glacier during the Late Pleistocene, and scoured deep within the Cretaceous limestones of the Pre-Alps. Holocene deltaic sediments form the lake-head delta, and deltas on east and west banks in mid-lake. A superb overview may be obtained from the Col de la Forclaz above the eastern side of the Annecey valley. Analysis of the sediments from Lac d'Annecey has allowed the identification of the late Pleistocene to Holocene sequence of environmental change in this Alpine area (Oldfield and Berthier 2001).

The Middle Rhône in Bugey and at Lyon From the Geneva lowland almost to Lyon the Rhône follows a course hemmed in to the north by the folds in the Jurassic rocks of the Jura and the Cretaceous rocks of the Pre-Alps primarily following Pleistocene glacial meltwater channels. A possible palaeochannel of an earlier Rhône runs through the Jura structures between Culoz and Ambérieu. It was abandoned during the last glaciation, with only locally derived younger sediments present. The modern highly engineered channel of the Rhône itself includes the power dam at Génissiat and diversionary channels near Belley. From Genix at the south end of the Belley ridge the channel heads northwest parallel with the Belley ridge. Downstream to Lagnieu (Fig. 13.6b), south of Ambérieu, the semi natural channel is on the threshold between meandering and anastomosing. From there to Lyon the channel is again heavily managed for power generation. Lyon itself marks the confluence of the Rhône and the Saône rivers. The Roman city of Lyon was west of the Saône, the medieval city and the modern downtown lie between the two rivers, and in the twentieth Century the city expanded east of the Rhône across late Pleistocene terraces. The view from the steps of Basilica Notre Dame at Fourvière, west of the Saône, encompasses these three elements of the city. South of the city between Givors and Vienne (another Roman Town) the River Rhône is within an incised valley, on whose slopes, especially southwards towards Condrieu, are Côtes du Rhône vineyards.

The Vanoise and Écrins Massifs (Fig. 13.4a; see also Fig. 4.4c) Both areas within the high "crystalline Alps "contain modern glaciers, and also late Pleistocene glaciated terrain. The Vanoise massif, south of Mont Blanc and lying between the upper Isère and Maurienne valleys, is accessible around its northeastern margins through the Tignes and Val d'Isère ski resorts and around its northwestern margins through the Courchevel/Val Thorens ski resorts. Access from the Maurienne valley to the south of the Vanoise is limited. Within the Vanoise massif itself are a few small (shrinking!) modern glaciers, and extensive late Pleistocene glacial forms. There are also considerable areas of high elevation Holocene gullying together with screes.

Further south the Écrins massif can be accessed from the Maurienne valley via one of a series of cols across the intervening ranges (Col de la Croix de Fer, leading to the Col de Glandon, on the D926 and D526 roads between St Jean-de-Maurienne and Bourg-d'Oisans). The Écrins can also be accessed via the Col du Galibier on the D902 road from St Michel-de-Maurienne to the Col de Lauteret (itself on the N91 road towards Briançon; Fig. 13.4a). There are numerous points of access into the Écrins massif itself, from the Briançon/Argentière area in the east, the Alpe-d'Huez/Deux Alpes area in the northwest, and the Corps/St Firmin area in the southwest. Perhaps the most spectacular area is the central area of the mountains around Mont Pelvoux and the upper Vallouise valley. There are Late Pleistocene-Holocene glacial landforms including erosional forms such as cirques and troughs, and Holocene depositional forms including Little Ice Age moraines. There are modern glaciers, albeit shrinking fast, slope and fluvial forms, mountain-slope gully systems, scree cones, debris cones and alluvial cones, and braided river channels.

The Grésivaudin/Chartreuse Including Mont Granier (see Hobléa, Chap. 18 in Fort and André, Eds. 2014) The high crystalline Alps are separated from the Cretaceous limestone Pre-Alps by a major structurally controlled depression aligned with an important strike/slip fault system along the Grésivaudin depression between Albertville and Grenoble, then south of Grenoble along the valley of the Drac. The crystalline Alps lie to the southeast of the Grésivaudin, with a series of Cretaceous limestone massifs to the northwest, each separated from the next by a major gap. The massifs are: the Bauges east of the Chambéry gap; then the Chartreuse; then south of Grenoble the Vercours. During the Pleistocene the Grésivaudin itself was occupied by glacial ice which spilled over and through to the west towards Lyon. Since deglaciation most of the depression has been occupied by the River Isère sourced within the crystalline Alps, and escaping northwest through the Grenoble gap into the southern part of the Lyon lowland. On the southeast corner of the Chartreuse Massif (opposite the town of Pontebarra on the floor of the Grésivaudin) is Mont Granier on which there was a major slope failure in 1248 AD causing a massive landslide (see Fig. 13.5b) towards the Grésivaudin.

At Grenoble the Isère turns through the gap between the Chartreuse and the Vercors massifs into the Lyon/Rhône lowland. During the last glaciation this lowland was occupied by various ice streams. A particularly pronounced former outlet of the Isère drainage runs through the centre of this lowland through Beaurepaire. The postglacial Isère hugs the southeastern perimeter of this lowland, along the western margin of the Vercors towards the Rhône near Valance.

The Drive Through the Vercors to Die Then Through the Drôme Valley (Fig. 13.5c, d) From Grenoble head south towards Villard-de-Lans within the north-south structures in the Jurassic limestones of the Vercors. From there head southwest through the limestone gorges of the Gorges de la Borne, then south along the strike of the Jurassic limestone ridges. Note the high scarp slopes to your left (east). Beyond the hairpins of the Col du Roussel you descend into the Drôme valley at

Die. From here follow the Drôme upstream through Luc en Diois to Le Claps. Here the Drôme undermined the Jurassic limestones, causing a major Holocene landslip, which partially blocked the valley (Fig. 13.5c). Note that in this area much of the Upper Jurassic limestone, characteristic of the area to the north, is replaced by black marls. Those marls plus the drier climate give the landscape a gullied semi-arid appearance characteristic of the Provençal Alps to the south (see Chap. 14).

Chapter 14
The Southern French Alps, Provence and the Mediterranean Coast

The Geology of the Southern French Alps and Languedoc

The basic structure of the southern French Alps (Fig. 14.1 map) resembles that of the northern French Alps, in that crystalline massifs line the Italian border, and they are succeeded westwards by Pre-Alpine massifs dominantly of Mesozoic sedimentary rocks. However, there are differences. South of the Écrins and Queyras crystalline massifs that mark the southernmost of the northern massifs (see Chap. 13), there are only two crystalline massifs within the southern part of the Alps. These are both lower in elevation than many in the north. They are the small Ubaye massif on the border between the northern and southern Alps, and further south the Mercantour massif. To the west of the Mercantour Massif is a domal structure exposing Triassic marls at its core. The potential geomorphic implications are discussed in the next section.

Other geological contrasts with the north relate to the Pre-Alps. In the south the Pre-Alpine belt of massifs formed of Mesozoic sedimentary rocks is much wider and more complex than that in the north. There are also lithological differences. The Lower Cretaceous Urgonian reef limestone is much less important in the south. Indeed, except in the far south, Cretaceous reef limestones as a whole are less important than in the north. On the other hand within the Upper Jurassic sequence there is an important black marl which, in the Mediterranean climate of the southern French Alps, gives rise to the dissected "badland" type landscapes of the "Terres Noires" (see Fig. 4.3a).

There are also major structural contrasts with the northern Alps. In the north the main structural trend of the Pre-Alpine massifs is NNE-SSW (see Fig. 13.1 map), parallel with the front of the crystalline massifs. To the south of the Vercors massif the strike begins to curve around to the southeast around the Écrins massif. Further south still the trend within the southern Pre-Alpine ranges is east-west (Fig. 14.1 map), affected by the interaction between the northern (Alpine) trend and the

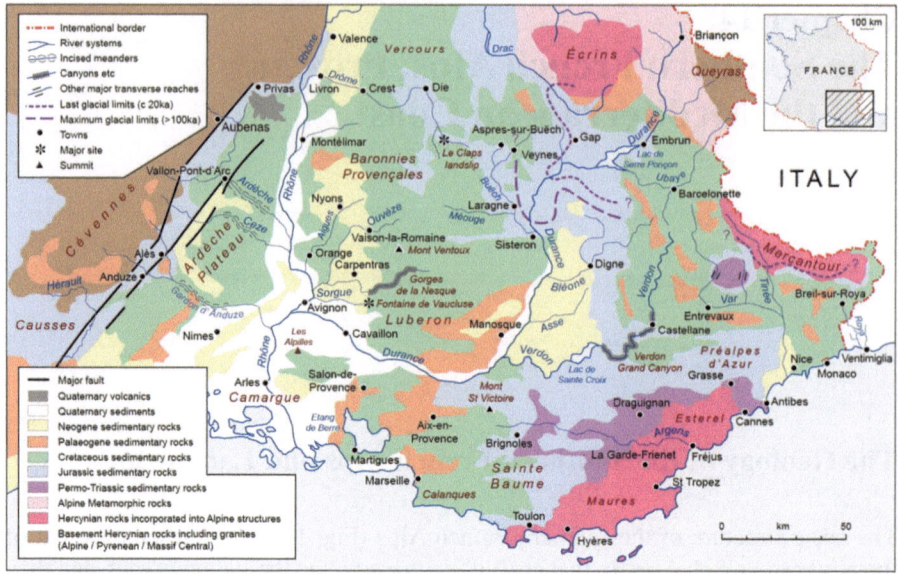

Fig. 14.1 Map of the southern French Alps and Provence

east-west (Pyrenean) trend (see Chap. 11). Thus the trend of the main Pre-Alpine ranges in Provence (those from around Sisteron southwards, dominantly of Lower Cretaceous rocks) is clearly east-west (Fig. 14.1 map), for example in Mont Ventoux, Montagne du Luberon, and Montagne Ste-Victoire ("Cézanne's mountain", near Aix-en-Provence, see Fig. 1.3d).

The Lower Cretaceous rocks, dominantly the Urgonian limestone, also form the spectacular Ardèche plateau on the west side of the Rhône valley (Figs. 14.1 map). This plateau is faulted below the Cévennes and the southern Massif Central (see Chap. 8), and stands to the north of the dominantly Cenozoic Languedoc coastal plain (see below).

East of the Languedoc coastal plain and the Camargue (the modern delta of the Rhône) coastal Provence comprises a predominantly erosional coast within which three geological zones can be identified: (i) East of Antibes (Fig. 14.1 map) extending beyond the Italian border is the coastal zone of the southern French Alps comprising folded Jurassic and Cretaceous limestones; (ii) from Antibes westwards to near Toulon the coastal zone comprises the Maures and Esterel massifs of Hercynian granites and metamorphic rocks structurally aligned with the E-W Pyrenean structural trend (Fig. 14.1 map). These rocks have affinities with the granitic rocks of western Corsica (see Chap. 15). They are flanked on their northern side by a belt of Permo-Triassic marls and sandstones, then by the Jurassic and Cretaceous rocks of central Provence; (iii) From Toulon west to Martigues, beyond Marseille and almost to the Camargue the geology is dominated by E-W folded Cretaceous limestones which locally incorporate Eocene sediments (Fig. 14.1 map).

The Camargue is the modern delta of the Rhone. The delta surface comprises modern fluvial and marine sediments dating from the period of modern sea level (c7ka), but burying late Pleistocene sediments. Adjacent to the delta (to the NE) is the Pleistocene Crau gravel plain, the former terminal fan of the Durance river. (The modern Durance joins the Rhône just south of Avignon, upstream from the apex of the modern delta.)

The Languedoc coastal plain (see later) is underlain mostly by Neogene and Quaternary coastal sediments, with the occasional hill unit of Cretaceous rocks, plus the Quaternary volcanic neck at Agde. The coastal plain is bounded to the north by the Cretaceous rocks of the Ardèche plateau, and to the west behind Minervois by the Hercynian metamorphics of the Montagne Noire. Lower Cretaceous rocks more or less reach the coast in the Montagne de la Clape, east of Narbonne. South and west of there is the small Rousillon lowland, before the Pyrenean rocks are encountered at Port-Vendres.

The Geomorphology of the Southern French Alps and Languedoc

Relief and Elevation There are relief contrasts with the northern Alps. First, the area occupied by crystalline massifs is much smaller, and on the whole elevations are lower. There are only two such massifs in the southern Alps, the small Ubaye massif on the border between north and south (maximum elevation at Aiguille de Chambeyron 3412 m), and the southernmost, Mercantour massif (maximum elevation at Cime du Gélas 3143 m) right on the Italian border.

Second, other contrasts relate to the Pre-Alps and other upland massifs. The total area occupied by such terrain is much larger in the south than in the north, but the definition of individual massifs is much less clear. To the west of the Rhône is the Ardèche plateau (maximum elevation 667 m) of lower Cretaceous limestone (see above) deeply trenched by the canyons of the Ardèche (Fig. 14.2a, b). In the coastal areas there are contrasts between the dominantly cliffed coasts of the majority of Provence and the depositional coasts from the Camargue westwards.

Fig. 14.2 West of the Rhône, the Ardèche plateau. (**a**) The natural arch at Vallon-Pont-d'Arc on the River Ardèche: created by an incised meander-cutoff. The incised meander cuts into the Lower Cretaceous Urgonian limestone: Shutterstock image 137145254 (*copyright: AJancso, Shutterstock 137145254*). (**b**) Further downstream on the Ardèche, incised meanders in the Ardèche gorges

East of the Rhône but west of the Durance system (Fig. 14.1 map) the terrain includes several rather loosely defined units. To the south of the Vercors massif and the Drôme valley (see Chap. 13, see also Fig. 13.5c) is Diois, a mountain area dominated by Lower Cretaceous massive limestones (which rise to 1570 m at La Berche). Further south, into Les Baronnies, the limestones thin and reach only lower elevations (1461 m at Montagne de Tuen), and there is more lowland terrain including badlands developed on the underlying Upper Jurassic marls (the 'Marnes Noires', Fig. 14.3a). Overall however, the relief is dominated by the folded Lower Cretaceous

Fig. 14.3 East of the Rhône, Les Baronnies and south to Mont Ventoux. (See also Figs. 1.3d Mont-St. Victoire, 4.3a Ste. Jalle badlands, 4.3d Periglacial patterned ground on Mont Ventoux). (**a**) Rills and gullies cut into the Lower Jurassic "Marnes Noires" badlands of the eastern Baronnies, near Ste. Jalle. (**b**) The outfacing escarpments marking the synclinal fold in Lower Cretaceous limestone in the Dentelles at Suzette, south of Vaison-la-Romaine. (**c**) Mont Ventoux a spectacular mountain reaching 1828 m, the highest point in western Provence. (What is more you can drive to the summit!) It is an escarpment of thrust-northward Lower Cretaceous Urgonian (reef) limestones

limestones, in both inverted and resequent relief configurations, with the various limestone bands forming escarpments (Fig. 14.3b).

Further south the topography is dominated by individual mountains: in Vaucluse by the isolated mountains of Mont Ventoux (1809 m, Fig. 14.3c); the Montagne du Luberon (1125 m); and to the southwest by Les Alpilles (which reach only 382 m). South of the Durance are Mont Ste. Victoire (1011 m, see Fig. 1.3d) and Massif de la Ste. Baume (1147 m).

To the northeast, the Pre-Alpine massifs also dominated by Lower Cretaceous limestones, include: Dévoluy to the northwest of Gap (reaching 2758 m at Grand Ferrand); the Montagne de Céüse (2016 m) within the Bochaine Range, southwest of Gap; and Les Monges (2115 m) in the Pre-Alps de Digne, south of Gap. Further east are the following Pre-Alpine massifs (Fig. 14.1 map): Embrunais (maximum elevation 2988 m at Parpaillon); then south of Barcelonnette the Trois-Évêchés massif (2961 m); and the Pelat massif (3051 m). Further southeast still are the Pre-Alps de Castellane (which reach 1930 m at the Mourre de Chanier within the Parc Géologique de Haute Provence), and the Pre-Alps de Nice (which reach 1778 m at Cime du Cheiron).

On the highest peaks within the Pre-Alps, for example Mont Ventoux (easily accessible by car), Pleistocene periglacial features such as well developed patterned ground are evident (see Fig. 4.3d). The Upper Jurassic rocks, especially the erosion-prone "Terres Noires", are characteristic of the lowlands and valley systems (Fig. 14.3a, see also Fig. 4.3a). Along the coast in the south, are the "Pyrenean" blocks of Hercynian metamorphic rocks and granites, the Maures Massif (maximum elevation 779 m at La Sauvette, Fig. 14.8b) and the Esterel Massif (maximum elevation 614 m at Mont Vinaigre).

Pleistocene Glaciation During the last glacial maximum (c20 ka BP) glaciers from the Écrins massif fed into the upper Durance valley near Briançon (Cossart et al. 2008), then down the valley, terminating somewhere short of Sisteron (somewhat short of the limit during the penultimate glaciation c150 ka BP, see Figs. 3.6 map, 13.2 map). Apart from a few patches of glacial moraine, there is little obvious glacial legacy within the middle part of the Durance valley. Very small remnant ice patches remain today within the Écrins (see Chap. 13). During the last glacial maximum glaciers were also present within the Mercantour massif. They were more extensive on the Italian (Argentera) side of the range. On the French side they did not extend far from the mountain front. However, they have left a wealth of glacial landforms within the Mercantour massif itself. These include erosional forms (cirques and troughs, many forming post-glacial lake basins), and depositional forms (moraines) within the main valleys. There were still small glaciers present in the Mercanour massif until the early twentieth Century, but there are none there today. There are also many mostly Pleistocene periglacial features (including frost shattered peaks, screes, and rock glaciers) formed both above the upper surface of the Pleistocene glacial ice and at lower elevations after ice decay.

Drainage, the Rhône and Durance Systems Three sets of river systems drain the southern French Alps and adjacent areas (Fig. 14.1 map). In the west is the lower Rhône system and its tributaries. In the centre is the Durance system, itself tributary to the seaward end of the lower Rhône (see below, next section of this Chapter). Along the coast southeast of the Durance system is a series of shorter coastal river systems (see next section).

The Lower Rhône System and Ardèche Plateau The lower Rhône River between Livron and Tarascon (Fig. 14.1 map) at the head of the Camargue delta (see next section of this Chapter) flows through a series of small basins partly confined by Lower Cretaceous limestone uplands. Although the modern channel of the Rhône has been radically modified throughout, within the wider sections of valley floor through the small basins there are often traces of former sinuous palaeochannels. The main modifications to the channel include floodwalls, weirs and by-pass channels in relation to hydro-electric plants, and diversions to provide cooling water for nuclear power plants.

The main tributaries from the west draining the Ardèche plateau (Figs. 14.1 map, 14.2) include the Ardèche itself and the Cèze. The Ardèche rises within the Cévennes, and crosses the main fault zone bounding the east side of the Massif Central (see Chap. 8). It then crosses the uplifted Cretaceous limestones of the Ardèche Plateau in a series of spectacular incised meanders (Fig. 14.2b) including the cutoff-related natural arch at Pont-d'Arc (Fig. 14.2a), south of Vallon-Pont-d'Arc. Within that area too are numerous limestone caves. South of the Ardèche, the Cèze has a similar, though less spectacular, course.

East of the Rhône, the Baronnies Provençales To the east of the Rhône, the River Drôme (already described in Chap. 13) really marks the boundary between the northern and the southern French Alps (Fig. 14.1 map). Its upper valley, together with that of its tributary the Maravel, has a very "southern" appearance. These valleys are floored by the erodible Upper Jurassic black shales giving rise to "Terres Noires" gullied landscapes (Fig. 14.3a, see also Fig. 4.3a). Away from headwater ravines the stream channels in this area tend towards braiding, but over the last 20 years or so, with the abandonment of some agricultural land and its replacement by forest, sediment supply has been reduced, and there has been a decrease in braiding (see Chap. 4 and references therein).

South of Diois two small tributaries of the Rhône are the Rubion and the Jabron (confluent with the Rhône at Montélimar). Both are small gravel-bed streams which drain the southern part of the Diois massif. Much larger are the next two main Rhône tributaries to the south: the Aigues/Eygues (spelt both ways) draining the boundary area between the Diois and Baronnies massifs; and the Ouvèze draining the northern Baronnies. Both rivers rise in upland terrain dominated by folded Cretaceous rocks. Both have reaches transverse to structure, especially near the mountain margins. This is probably due to superimposition from an earlier Cenozoic cover, rather than to antecedence (see Chap. 11 for discussion of similar features on the margins of the Pyrenees). Within these transverse reaches the valleys are often incised meandering valleys. However the modern gravel-bed channels are dominated by braiding rather than by meandering. South of Dieulefit (a lovely name: "God made it") the upper reaches of both rivers tend to be structurally aligned along east-west orientated anticlines within Cretaceous limestones. The anticlines are cored by Upper Jurassic black marls, which are prone to gully and badland development ("Terres Noires"). Gullied terrain is particularly well developed in the Buis-les-Baronnies and Ste. Jalle areas (Fig. 14.3a, see also Fig. 4.3a). The terrain also picks out the east-west folded Cretaceous limestones further south particularly at Suzette, south of Vaison-la-Romaine (Fig. 14.3b). Just southeast of Vaison-la-Romaine is the thrust-forward mass of Mont Ventoux (Fig. 14.3c), the highest point in the Baronnies with spectacular views plus evidence of Pleistocene periglacial activity (see Fig. 4.3d).

Further south is the karstic plateau of Vaucluse drained by the Nesque, a tributary of the Ouvèze (note the Gorges de la Nesque), and the Sorgue (fed by the enormous karstic springs at the Fontaine-de-Vaucluse). South of there is the east-west Luberon range, south of which is the Lower Durance valley. South of the Durance are more east-west karstic limestone ridges, including in the west Les Alpilles at Les Beaux, an archaeological site, and further east near Aix-en-Provence, Montagne Ste. Victoire (see Fig. 1.3d).

The main left-bank tributary of the Lower Rhône, with its confluence just south of Avignon, is the Durance (see below). Further downstream, with its confluence with the Rhône near Tarascon, is the right-bank tributary the Gard/Gardon. This river rises in the west within the Cévennes and crosses the fragmented southern part of the Ardèche plateau, here dominantly of Cretaceous limestones. Through these limestones it has cut incised meandering canyons upstream of the famous Roman bridge at Pont du Gard. From this point on, the Rhône enters the Camargue delta. I deal with the geomorphology of that area later in this chapter.

The Durance System, Central Provence   The second main river system draining this area is the Durance (Fig. 14.1 map). This river rises in the High Alps on the boundary between the northern and southern French Alps, and is fed from the crystalline massifs of Les Écrins, Queyras and Ubaye. Within most of the upper Durance valley, around Briançon and downstream as far as L'Argentière, the valley is narrow and the channel is mostly cut into bedrock. Further downstream are Late Pleistocene river terraces (see Fig. 3.7c). Through this reach braiding dominates. Beyond Embrun is the Serre Ponçon reservoir created by the enormous 123 m-high earth-cored dam. An indication of the high sediment supply from the Durance into the reservoir is given by the amount of accretion at the head of the lake that has accumulated since the dam was completed in 1960. In addition there is a huge fan delta fed by the Torrent de Boscodon on the south side of the lake. The southern arm of the lake is fed by the River Ubaye that drains the mountain area to the east between the Ubaye and Mercantour crystalline massifs. The river itself has mostly a steep single-thread channel locally incised into bedrock, but there is braiding in the upstream reaches near Barcelonette.

Southwards, as the climate becomes increasingly Mediterranean (hot/dry summers, often with heavy autumn/winter storms), natural (and human-disturbed) vegetation is thinner, making the soils, and especially those on the upper Jurassic shale bedrock, prone to erosion. There are many gully and badland areas within the middle Durance valley (Fig. 14.4). For most of the way from the Serre Ponçon dam downstream to Sisteron, the main flow of the Durance is in an artificial channel, supplying hydro-electric power plants. In many places, for example downstream of Thèze (at around the last glacial limit, Fig. 14.4), the old channel is preserved on the valley floor as a meandering palaeochannel below a suite of late Quaternary river terraces. Within that channel there is a micro-pattern of braid bars related to the now much lower modern flow volumes. Just upstream of Tallard is another interesting section of the valley. The valley sides are cut in Upper Jurassic dark shales with well developed "Terres Noires" badlands and gully systems. The valley floor has a clear sequence of Late-Pleistocene to Holocene river terraces (Fig. 14.5a). Although the main flow is culverted through this reach, the natural channel carries sufficient flow to allow the development of a braided pattern of gravel bars within the overally meandering channel. Just short of Sisteron is a left-bank tributary, the Sasse, that drains the Jurassic limestone and shale terrain of the Digne pre-Alps. Within its drainage basin there are "Terres Noires" gullied and badland areas. For most of its course it has a braided channel.

Sisteron itself has an impressive site (Figs. 14.4 map, 14.5a). The River Durance cuts a gorge though near-vertical Jurassic limestone beds. Just upstream of this gorge is the right-bank tributary junction with the River Buëch. The River Buëch drains Pre-Alpine terrain in the Diois and Baronnies massifs, which are dominated by lower Cretaceous limestones (Fig. 14.5b) with an east-west structural grain. Within the valleys the underlying Upper Jurassic shales are exposed, leading to classic "Terres Noires" badland and gully landscapes. I have always regarded the lower Buëch valley, between Montrond and Sisteron but centred on Laragne, as an excellent area to see this type of upper Provençal landscape (Figs. 14.4 map, 14.5b).

Downstream of Laragne the Buëch itself has a superbly developed braided channel pattern (see Fig. 4.5b). By way of variety, nearby (southwest of Laragne) is the beautiful incised meandering gorge of the Méouge (Fig. 14.5c). Furthermore, in several areas, both south and east of Laragne there are excellent "Terres Noires" badlands and gully systems.

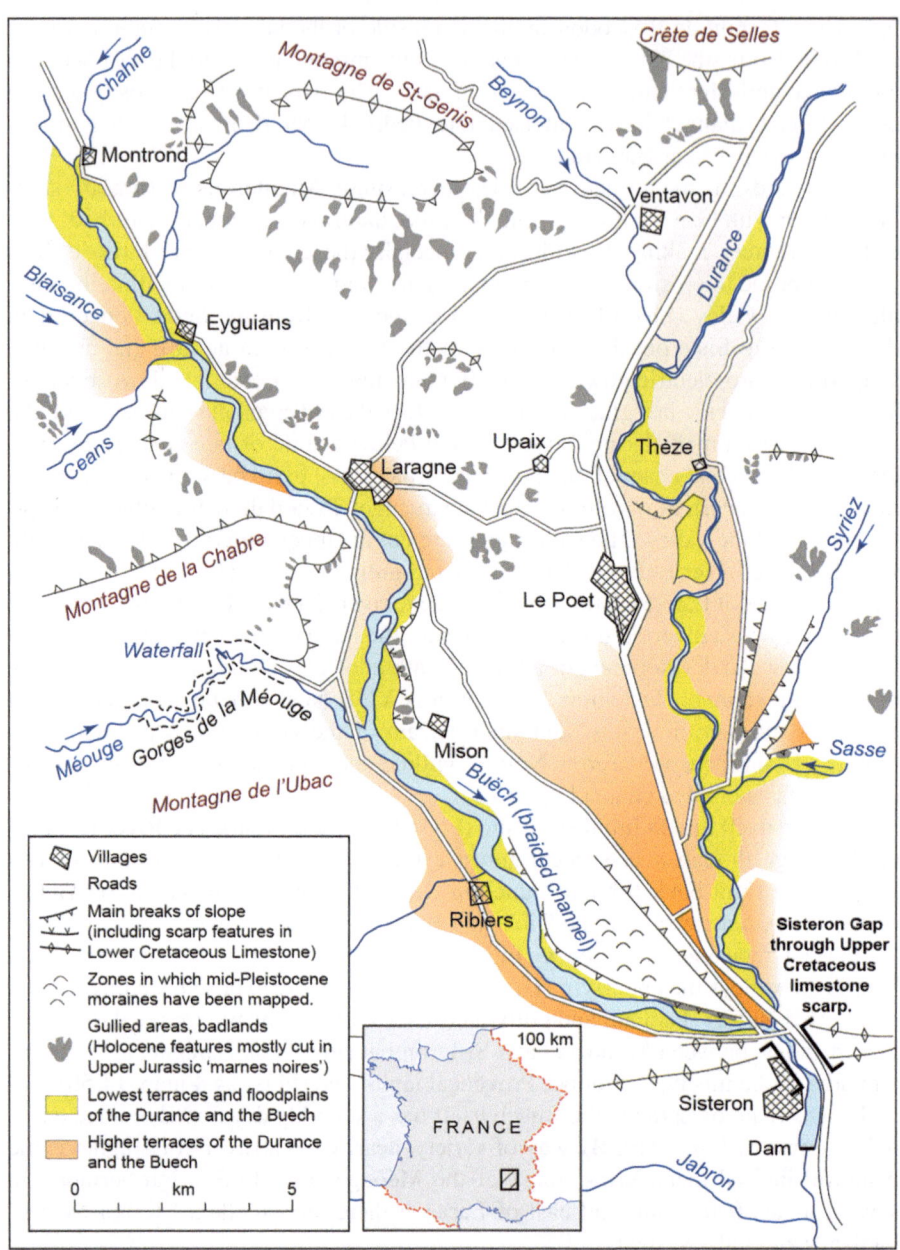

Fig. 14.4 Map of the Durance/Buëch valleys in the Laragne/Sisteron area

Fig. 14.5 The Durance/Buëch valleys and surrounding mountain areas. (**a**) The middle Durance valley at Sisteron, looking upstream to where the basin is cut within Upper Jurassic dominantly shaley rocks, and where the Durance River is joined by the Buëch (from the west, left of the photo). Upstream of the confluence the Buëch has dominantly a braided channel (see Fig. 4.5b). Pleistocene river terraces are evident on both sides of the Durance (as those shown on Fig. 3.7c). In the distance are the Cretaceous limestone mountains of the Montagne de St Genis (see Fig. 14.4 map). (**b**) The terrain northeast of Laragne, looking north towards the Lower Cretaceous limestone synclinal mountain of St Genis. The (locally gullied) subdued terrain in the middle distance is developed on Upper Jurassic marls (see Fig. 14.4 map). (**c**) The Gorge de la Méouge southwest of Laragne, cut into Lower Cretaceous limestone. Note the bedrock channel of the Méouge (see map Fig. 14.4)

South of Sisteron, the Durance and Its Tributaries South from Sisteron, down-stream to the Verdon confluence near Mirabeau, the Durance crosses the "Tertiary" sedimentary basin of central Provence. There are slightly deformed Oligocene and Miocene sediments to the west and almost undeformed upper Miocene and Pliocene detrital sediments forming the Valensole plateau to the east (Fig. 14.1 map). These sediments are locally undercut, displaying "hoodoos", erosional pillars (Fig. 14.6a). The Durance valley itself includes late Quaternary river terraces and a few (now stabilised) tributary junction alluvial fans (more obvious on air photos and satellite images than on the ground). The river, where it is at least semi-natural, has a braided channel but elsewhere has been highly modified by weirs and dams with most of the flow channelized into by-pass channels (again mostly for the purpose of power generation or nuclear power station cooling). The former channel is also obvious, only carrying significant flow amounts during floods, but preserving a gravel bed, often in the form of braids. Braiding persists throughout the middle and lower reaches of the Durance.

Fig. 14.6 The Durance valley downstream from Sisteron, and its eastern tributaries. (**a**) The "Hoodoos": Erosional pillars cut in Miocene marls, on the margins of the Durance valley at Les Mées just below the Bléone confluence downstream of Sisteron. (**b**) The braided channel of the River Asse. (**c**) The major east-bank tributary of the Durance is the Verdon: This river joins the main river south of Manosque. It is deeply trenched into the karstic limestone plateau formed on Upper Jurassic limestones. Here is shown a view of the Grand Canyon of the Verdon looking west (downstream), below Pointe Sublime (see also Fig. 4.4b)

There are three major left-bank tributaries to the Durance in this area, the Bléone, the Asse and the Verdon (Fig. 14.1 map). The Bléone rises within the rugged, deeply dissected terrain developed on a variety of Cretaceous rocks within the Pre-Alpine Trois Évèchés massif. There is much active modern slope erosion in this area, feeding sediment into the braided channel of the upper Bléone. At Digne the river crosses the Pre-Alpine mountain front. Downstream of Digne, the valley is bounded both to the north and the south by the dissected terrain of the northern part of the central Provence sedimentary basin. This "mid-Tertiary" basin is developed here in Miocene rocks. The valley itself has late Pleistocene to Holocene river terraces, below which is the modern braided channel.

Further south, the Asse (Fig. 14.6b) has similar characteristics, rising in the complex dissected terrain of the Trois Évèchés and Castellane Pre-Alps in folded Jurassic and Cretaceous rocks. Its upper course is followed by the Digne-to-Nice "Route Napoléon", as well as by the narrow-gauge Chemin de Fer de Provence. For most of this course the channel of the Asse is braided, albeit bedrock-confined in places, including the spectacular transverse gorge just upstream of Châteauredon. Beyond Bras-d'Asse it flows through the sedimentary basin with dissected Miocene sediments to the north, and the virtually undissected Pliocene sediments capped by red soils of the Valensole plateau to the south. Within the valley floor of the lower Asse the channel maintains a braided pattern locally below late Quaternary river terraces. There are superb views of this landform assemblage from the minor road northeast of Valensole.

The third, and most spectacular of these east-bank Durance tributaries is the Verdon (Figs. 14.1 map and 14.6c). It rises near Allos, and is fed by sources within the Mont Pelat range. This range, although formed dominantly of Upper Cretaceous rocks rather than of crystalline rocks, has high relief reaching elevations in excess of 3000 m. It has an "alpine" appearance having been glaciated during the Late Pleistocene, creating a landscape of cirques and other glacial features. Elsewhere the upper basin of the Verdon is developed in Upper Jurassic and Lower Cretaceous rocks. The terrain is deeply dissected and in many places erosion is active, evidenced by screes, gully systems and fresh tributary-junction alluvial fans. Where the Verdon valley floor widens the channel is braided, again reflecting the high rates of sediment supply. Otherwise it is a single rock-cut channel constrained by bedrock.

The Castellane area, in the middle reaches of the Verdon valley, marks an important structural change. To the north the main fold structures, affecting especially the Cretaceous rocks and therefore the topographic ridges, are aligned north-south with the Alpine trend. Near Castellane they swing round to an east-west alignment to accord with the Provençale/Pyrenean trend.

Near Castellane too there are several dams on the Verdon creating artificial lakes. There are two dams upstream of Castellane, the upper one creating a large lake as far upstream as St André-les-Alpes. Downstream of Castellane is the Grand Canyon of the Verdon (Fig. 14.6c, see also Fig. 4.4b), a truly spectacular geomorphological site, with many viewpoints along the roads above the canyon. Perhaps the best known is Pointe Sublime. The canyon is cut mainly in Jurassic limestones, but immediately downstream of the canyon, just beyond Moustiers-Ste Marie, the river lies between the (Pliocene) Valensole Plateau to the north and Jurassic limestone terrain to the south. Moustiers-Ste Marie lies at the head of another artificial lake, the Lac de Ste. Croix. There are several more dams and lake sections, both upstream and downstream of the Basses Gorges du Verdon, before the confluence of the Verdon with the Durance near Avignon. Otherwise, where it gets the chance between dammed reaches with lakes and the bedrock canyons, the Verdon has a tendency to braid.

The whole area between the Bléone and the Verdon is conserved not only as the "Parc Naturel Regional du Verdon", but also as the "Reserve Naturelle Géologique de Haute Provence". This designation marks its significance for both geology and geomorphology (see Chap. 16).

Near Mirabeau the Durance valley is confined between Cretaceous limestone uplands, but the valley opens out from there downstream to near Cavaillon. A few kilometres short of Cavaillon a former course of the Durance can be traced southwestwards towards the Crau gravel plain, itself a former (Pleistocene) fan of the Durance (see later in this Chapter). Just short of Cavaillon there is another narrowing below the western end of the Montagne du Luberon and the Alpilles chain, both developed in Cretaceous limestone. Presumably this gap was incised in response to lower base levels during the Pleistocene. Throughout this reach the valley form is similar to that described above (terraces, a by-pass channel taking most of the flow, gravel braids within the former natural channel). However, near Cadenet, and similarly near Mallemort, EDF by-pass channels carry most of the flow of the Durance southwards (following the Pleistocene course of the Durance towards the Crau and the Étang de Berre). From Mallemort downstream through Cavaillon and beyond, the post-Pleistocene Durance has a semi natural braided gravel-bed channel, but it has lost most of its discharge. Beyond Cavaillon to the confluence with the Rhône just south of Avignon, the terrain opens out and the channel retains a braided form, although it carries very little flow.

Drainage, the Far Southeast: Inland Areas In the far southeast is the famous Riviera coast, the Côte d'Azur, extending from Cannes through Nice to the Italian border (see below). The coast is more or less built up all the way from Cannes to the Italian border. However, I do not suppose that many people come here to study the geomorphology—sun, sea, sand, partying perhaps! For some real geomorphology—head inland!

The Mercantour Massif and the Alpes Maritimes Inland from the Côte d'Azur are spectacular mountain landscapes culminating in the crystalline Alpine Mercantour massif (Fig. 14.7a). The highest mountains of the Mercantour massif just reach elevations of 3000 m. They were intensively glaciated during the Pleistocene. There are no modern glaciers, but there is a wealth of fresh glacial landforms; cirques, troughs, moraines, plus deeply incised scree-mantled valley-side slopes. The glacial forms are exceptionally well developed on the Italian side of the border (in the so-called Argentera massif), but there is one area on the French side of the border where glacial landforms are well developed, the Mont Bégo area, also known as the Vallée des Merveilles (see Magail and Simon, Chap. 21 in Fort and André, Eds. 2014). This area has the added attraction that it is famous for its Bronze-Age petro-glyphs. It is difficult to access. The road network as a whole in the Mercantour is not good, especially if you wish to access the Italian side of the border. For example, there is a tunnel under the Col de Tende at the eastern end of the range, but if you actually wish to cross the pass itself, there are 42 hairpin bends on the single-track road to the summit.

South of the Mercantour massif are the Alpes Maritimes Pre-Alps, developed in folded Cretaceous limestones, often capped by Eocene sediments. The Pre-Alps have maximum elevations up to around 2000 m in the north, but only up to about 1000 m nearer the coast. The steep regional gradient is reflected in the deep incision of the drainage network. There are two main drainages, the Roya in the east reaching the sea at Ventimiglia just over the Italian border, and the Var in the west reaching the sea at Nice.

Fig. 14.7 Massif de Mercantour and the Var valley. (**a**) Deeply incised headwater valley within the Mercantour Massif, near Isola. Note the scree-mantled lower slope below the bedrock midslope on the right. (**b**) Landsat image of the Var valley near Entrevaux [E 6.47' N 43.57']. Note the west-east alignment of the Var valley downstream (east) of Entrevaux. This alignment parallels the strike of the Cretaceous limestone. Note also the domal structure (reddish terrain—Triassic rocks), north-east of Enrevaux and the transverse reaches upstream (west) of Entrevaux. (**c**) The village of Entrevaux: perched above the incised channel of the River Var: Shutterstock image 536887603 *(copyright: Allard One, Shutterstock 536887603)*

The Roya   The Roya rises on the French side of the Col de Tende, then picks up deeply incised tributaries from the west from the western side of the Mercantour massif. Downstream from the col., the Roya has a bedrock channel in a deep incised valley through Tende to Saint Dalmas (near La Brigue—Note the multiple spiral tunnels on the railway!). At St Dalmas-deTende the Roya picks up a steep tributary from the west, the Minière, from the Vallée des Merveilles (see above). The steep incised bedrock channel continues through Saorge (via a transverse reach with incised meanders) to Breil-sur-Roya. This is where the valley road is joined by the direct road from Nice through the Sospel pass. Beyond Breil the Roya passes into Italy and through some tortuous incised meanders to Trucco from where the valley widens and the channel becomes braided downstream to Ventimiglia. En route it receives a west-bank tributary, the Bévera, another tortuous incised meandering bedrock channel, rising on the French side of the border near Sospel.

The Var   The other major drainage of the Alpes Maritimes is the Var (Fig. 14.7). The Var rises in central Provence to the west of the Mercantour massif, east of Mont Pelat and to the west of the source of its main tributary, the Tinée (see below). Both rivers are deeply incised (Fig. 14.7a). The mountains surrounding the source of the Var are formed of Cretaceous limestones, but the headwater valleys expose Upper Jurassic rocks. Near Chapel St Sébastien there are spectacular gullies and badlands cut into the "Terres Noires" shales. Not surprising with such high rates of erosion within the upper catchment, the upper Var has a sediment-rich channel, with sand and gravel bars and boulder berms. Just short of St Martin-d'Entraunes the Var picks up a major sediment-laden east-bank tributary that drains an intensely gullied catchment. From this point downstream the main river has a truly braided channel pattern, augmented by east- and west-bank sediment-rich tributaries which drain gullied catchments. Further on there are more sediment-rich tributaries. The high rates of sediment supply were also effective in the past, with a large tributary-junction alluvial fan deposited at the mouth of a steep east-bank tributary at Villeneuve-d'Entraunes. Further on still, just beyond Guillaumes, a domal structure brings up the underlying red Triassic sandstone (Fig. 14.7b). Is this domal structure diapiric, a salt dome or gypsum dome destabilised by Neogene tectonics? If so, has there been an influence on drainage development? I do not know the answer to those questions. Maybe there has been some French research on this topic of which I am unaware. Whatever!—I would like to get out there again!

The river, now with a steep bedrock channel, cuts through the northern margin of the dome in a spectacular incised canyon reach through the Gorges du Daluis. Beyond the gorges the river passes again onto Jurassic rocks (more gullying, and a braided channel again), then at Pont de Gueydan onto Cretaceous limestones. These have the now characteristic east-west fold and thrust fault structures of the southern Provençal Pre-Alps (Fig. 14.7b).

At Pont de Gueydan the Var is joined by the strike-orientated Coulomp valley. This gives access to the Vaire valley, and the Annot area (more railway spirals—the narrow gauge Chemin de Fer de Provence), mostly through Cretaceous rocks, and eventually to the upper reaches of the Verdon. Between Pont de Gueydan and Entrevaux is an extraordinary zone of drainage complexity where the Var cuts a transverse course (Fig. 14.7b) across both northern and southern limbs of an anti-cline in Cretaceous limestones. Many of the other rivers and streams in this area also cross the same east-west fold axes. I wonder how these patterns might relate to any diapiric activity in the domal structure described above.

At Entrevaux (a spectacular village site, Fig. 14.7c) the orientation of the Var valley changes abruptly to eastwards following the strike of the Cretaceous rocks. Throughout the reach to Puget-Théniers the river has a braided channel. From there the valley narrows and steepens. The channel is narrower, but still locally braids, especially where tributaries feed more sediment into the main river. This is the case of the Cians, a north-bank tributary whose junction is just upstream of Touët-sur-Var. The Cians is a steep transverse stream flowing in canyons across the domal structure in Triassic rocks, described above.

From this area, past Malaussène, the Var valley narrows and steepens. The narrow but still braided channel occupies virtually the whole valley floor. The river is now effectively in a canyon, the Défilé du Chaudan. In this reach the river picks up two important north-bank tributaries, the Tinée (see above) and the Vésubie, both of which rise in the Mercantour massif.

The Tinée rises in an area of complex geology high in the Mercantour massif in the St Étienne de-Tinée area between the main Mercantour range and the Pelat/ Mournier range to the southwest. The uppermost parts of the headwater catchments are highly active geomorphologically, with extensive shale badland and gullied erosional areas, plus depositional sites: scree cones and alluvial fans. The stream channels show a combination of gravel-bed partial braiding and incisional bedrock reaches. Downstream through St Étienne-de-Tinée to Isola the hillslopes are more stable and the valley floor has a partially braided channel within a floodplain. The tributary catchment of Isola 2000 (the ski area) is also eroded, some of which relates to the modern disturbance associated with the development of the ski resort. Downstream from Isola the Tinée becomes increasingly incised through the Gorges de Vallabres to the confluence with the Var.

The other main tributary to the Var in this area is the Vésubie, another left-bank tributary that rises in eroded headwater catchments in the main Mercantour range upstream of St Martin-Vésubie (scene of the devastating floods in October 2020). Downstream the valley of the Vésubie becomes increasingly incised through the Gorges de la Vésubie towards its confluence with the Var.

A little further down the Var at St Martin-du-Var the Var is joined by a right-bank tributary from the west, the Estéron. This river is a strike-orientated drainage within a dominantly limestone catchment within east-west folded and faulted Jurassic and Cretaceous rocks. Locally however, there are transverse reaches, where the valley crosses the strike of the terrain between two accordant reaches. One such transverse reach is south of Roquestéron. There are several others further upstream. The upper and middle reaches of the Estéron have single-thread alluvial channels, locally with minor braiding, but towards the confluence with the Var the channel is a bedrock channel within incised gorge reaches. The lowermost reach of the Estéron together with the lower Var are braided reaches, albeit confined by floodwalls.

An intriguing characteristic of the rivers in the Pre-Alpine area is the development of transverse reaches. These rivers are very similar to those of the Pyrenean foreland (see Chap. 11), largely strike-orientated rivers draining terrain comprising dominantly folded and faulted Cretaceous limestones rocks. Such rivers are locally transverse, "jumping" from one accordant reach to another across the structures. I do not know the origin for this. In neither case can glaciation be invoked as a possible cause. Could it be due to superimposition from an earlier unconformable cover rock? That is possible. In both areas the Cretaceous rocks have had a cap of Palaeogene rocks. Or could the transverse drainage alignments be antecedent, in other words pre-dating the last phase (Miocene?) of structural deformation? As both areas are dominantly limestone areas, could karstic processes be involved? In the case of the Var, I wonder about any relationship with any diapirism within the domal structure northeast of Entrevaux (Fig. 14.7b). I do not know the answer to these questions. I am not aware of published research on this topic. Maybe there is none; on the other hand, maybe I am not familiar enough with the French research literature.

The Provençal Coastal Area The three sections of the Provencal coastal zone, differentiated by their geologies (see above) are also differentiated on the basis of their geomorphologies. The eastern sector, the Côte d'Azur, is a steep mountain coast with little flat land near the coast. What flat land that exists is occupied primarily by town sites, particularly Nice and Cannes. The flat land of the Var delta is occupied by Nice Airport. There is only limited access to the coast itself, which tends to be of low cliffs with narrow beaches below of sand or shingle.

The central sector, Cannes to Toulon, is dominated by the Hercynian granites and metamorphic rocks of the Esterel and Maures (Pyrenean) massifs incorporated into the Alpine/Pyrenean structures. These massifs are flanked north and west by Triassic sandstones, especially in the Draguignan area. The relief is not great. Maximum elevations are less than 700 m in the Maures and just over 500 m in the Esterel. Locally inland the granitic rocks are exposed and there are tors on the higher more exposed surfaces (eg. at Mont Vinaigre in the Esterel massif). The attractive Esterel coast is dominated by low red granite cliffs (Fig. 14.8a) with mostly small bayhead beaches. In places the granite cliffs are bevelled by raised wave-cut platforms which cut across the structures in the underlying granite. They are presumably late Pleistocene in age, and date from an interglacial sea level higher than that of today. Inland the Maures are mostly wooded, but there are steep slopes and in places exposed bedrock (Fig. 14.8b). From south of Le Muy on the north side of the Maures, and from the viewpoint above La Garde-Freinet in the centre of the Maures, there are spectacular views northwards across the Triassic-based Argens valley towards the Jurassic limestone plateau south of the Verdon. The River Argens rises in the Jurassic limestone country south of the lower Verdon, then drains the Triassic sandstone country west of Draguignan. Its upper and middle course has incised misfit meanders. It then skirts the north of the Maures. From Le Muy to the sea at Fréjus the Argens has a stable meandering channel within a floodplain. Within the Maures massif itself the other short coastal streams are mostly ephemeral, with steep bedrock channels, similar to those in the Esterel massif.

Fig. 14.8 The Maures and Esterel massifs. (**a**) Esterel coastal cliffs, cut in Hercynian granites. (**b**) Rocky hillslopes in the Maures massif, near La Garde-Frienet, showing weathering features in the granitic bedrock

The Maures coast, like the Esterel coast, has low granite cliffs interspersed by small bayhead pocket beaches.

The third sector of the coast is that between Toulon and Marseille and beyond. The inland terrain is dominated by east-west ridges, near Toulon mostly of thrust-forward Jurassic limestones. Towards Marseille the relief is dominated by the Cretaceous limestones of the Massif de la Ste. Baume which rises to over 1000 m. Beyond Marseille the much lower Cretaceous limestone ridge, the Chaîne d l'Estaque (200 m), separates the open sea (including the Golfe de Fos) from the Étang de Berre. Beyond this area is the gravel plain of the Crau, the Pleistocene delta of the Durance, which leads towards the Camargue (the modern Rhône Delta). In the Marseille area the coast itself tends to be limestone-cliff dominated (eg. the Estaques cliffs west of Marseille, the Calanques cliffs and inlets east of Marseille). It is mostly urbanised with only the cliffed headlands retaining anything like natural landscapes.

The Rhone Delta (the Camargue) and the Languedoc Coast The Mediterranean coastal area including and west of the Camargue differs considerably from that to the east. The coast to the east is dominantly an erosional coast backed by usually low cliffs, with beach development mostly as pocket beaches between cliffed headlands (see above). The western coastal area is dominantly depositional (Fig. 14.9 map), apart from a few headlands, all the way from the Camargue to the eastern end of the Pyrenees (see Chap. 11). It falls naturally into two units: the Camargue (ie the Rhône delta, Fig. 14.10 map), and the Languedoc coastal plain (Fig. 14.9 map).

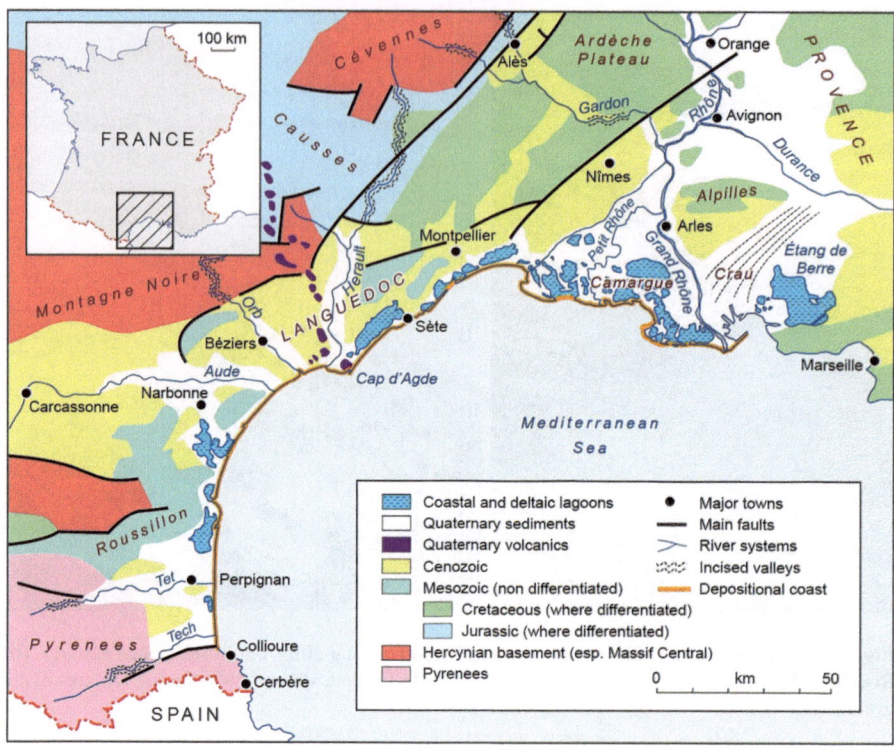

Fig. 14.9 Map of the geology and river systems of the Camargue/Languedoc area

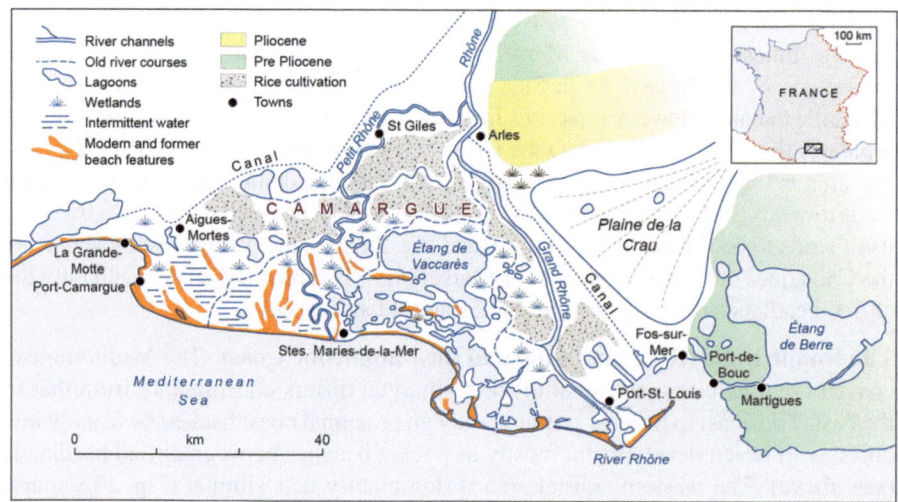

Fig. 14.10 Map of the Camargue

Fig. 14.11 The Camargue. Lagoons, marshland and flamingos: A Camargue landscape: étangs on the western margin of the Camargue near Aigues-Mortes. Shutterstock image 1124868272 (*copyright: Francesca 2011 X Shutterstock 1124868272*)

The Camargue Downstream from Avignon the lowermost Rhône, augmented from the west by the Gardon and from the east by the Durance, feeds into the head of the Rhône Delta, the Camargue. The Camargue (Figs. 14.10 map, 14.11) is an iconic landscape involving flamingos, black fighting bulls and wild horses, but those human associations depend very much on the physical characteristics of the area. The evolution of the delta area can be traced back to the late Miocene (Messinian) salinity crisis when the Mediterranean became landlocked (see Chap. 2), and sea levels fell dramatically through evaporation. The result was the entrenchment of the then Rhône river, forming a canyon on what is now the seabed offshore. When sea levels rose again a new delta formed during the Pliocene and early Pleistocene roughly where the present delta is. These sediments bound the landward side of the modern delta (Fig. 14.10 map) including the palaeo-Durance gravels that form the Crau plain to the east of the modern delta. During the Pleistocene the delta zone would have been trenched during "glacial" low sea levels, forming what are now submarine canyons offshore.

The geomorphology of the present delta has resulted from the interaction between fluvial and marine processes over the last 7000 years, the period since the post-glacial rise in sea level, plus human-induced processes in more recent years. The sudden reduction in gradient from the fluvial valley floor upstream of Tarascon promoted sedimentation, initially in the upper part of the delta. It also promoted channel avulsion (ie. channel splitting resulting from washovers during flood conditions). There are two main arms of the Rhône today, splitting at Arles into the Grand Rhône feeding the east of the delta, and the Petit Rhône feeding the west. The Grand Rhône is a wide mildly sinuous channel taking the bulk of the modern river flow. It has artificial flood protection banks on both sides of the channel. The Petit Rhône is a much smaller meandering channel. Both channels have been subject to previous avulsions. The present channel of the Grand Rhône through Port St Louis was an avulsion from an earlier channel to the southwest through the Bras de Fer. There had been other older channels into the central Camargue feeding the large lagoonal areas (Fig. 14.11). The present seaward exit of the Petit Rhône through Saintes-Maries-de-la-Mer was an avulsion from several earlier channels feeding the so-called Petite Camargue south of Aigues-Mortes. Lagoons (Fig. 14.11) have formed between the distributary arms. The largest lagoon is the "Vaccarès" in the centre of the delta. At the same time the shoreline has advanced seawards, south of the Vaccarès lagoon, leaving behind a series of beach ridges, which mark the stages in the overall growth of the delta. These culminate in the modern beach ridge. Longshore drift is from east to west so that these ridges have grown as spits in that direction. Since the nineteenth Century, reduction in the sediment loads carried by the Rhône, together with the channelization and embankment of the Grand Rhône, has reduced sediment delivery to the delta, with the result that rather than continued delta growth, the seaward face of the delta has been subject to marine erosion.

Languedoc, River Systems (Fig. 14.9 map): Extending southwest from the Camargue is the Languedoc coastal plain, backed by the Cévennes mountains (see Chap. 8, Massif Central). Crossing the coastal plain immediately to the west of the Camargue is the Vidourle, which drains the southern margin of the Cévennes and feeds into the western edge of the Camargue at Le Grau-du-Roi. Most notable are the large incised meanders in its middle course between Quissac and Sommières. The next major drainage to the west is the Hérault. This river rises in the Cévennes, picking up a right bank tributary, the Vis, from the west that drains the eastern part of the Causses area (see Chap. 8). The headwaters of the Hérault near Ganges include major incised transverse reaches across the east-west structures of the Cévennes, similar to the transverse courses of the streams draining north from the eastern Pyrenees (either superimposed or antecedent? see Chap. 11; see also Fig. 11.2 map). There are further incised meandering reaches in the middle reaches of the Hérault through the Jurassic limestone terrain to the north of Aniane. South of there, the sinuous to meandering channel is within a floodplain across the Languedoc coastal plain, reaching the sea at Agde.

Further west the Orb and its tributary the Mare rise in the broken terrain of the eastern Monts de Lacaune. These mountains are formed of Palaeozoic rocks that were incorporated within the Palaeogene Pyrenean thrust system. The headwaters of the Orb are dammed to form the Avène reservoir below which the river is in a floodplain within incised meanders. Locally, there are bedrock reaches and further downstream there are late Quaternary terrace remnants within the valley. Near Bédarieux the Orb is joined by the Mare, a wide meandering gravel-bed stream. From Bédarieux downstream to Tarassac the valley is fault-aligned along a major thrust fault bounding Palaeozoic metamorphic rocks to the north. Through this reach the river has a sinuous channel below late Quaternary terraces. There the Orb turns south to become transverse across east-west folded

Palaeozoic sedimentary rocks, then emerges onto the Languedoc coastal plain at Cessenon-sur-Orb from where it has a sinuous to meandering channel within a wide floodplain downstream to Béziers, reaching the sea at Valras-Plage. Draining the Carcassonne gate and much of southern Languedoc is the River Aude. This river rises in the eastern axial Pyrenees (see Chap. 11) south of Axat. It then becomes a transverse drainage across the folds and thrusts in the Cretaceous rocks of the Pyrenean frontal ranges through Quillan (see Chap. 11, see also Fig. 11.2 map) to reach the edge of the Carcassonne gate north of Limoux. At Carcassonne it turns east, into Languedoc, skirting the southern margins of Minervois, to reach the coast south of Valras-Plage. Much of the channel of the Aude, above Carcassonne is a single-thread channel within a narrow valley. From there downstream it has mostly a single-thread loosely meandering channel, with very occasional gravel islands, below Quaternary terraces. Its main tributary from the south is the Orbieu, which drains the uplands of Corbières, flowing through Lagrasse to join the Aude northwest of Narbonne. The Orbieu has mostly a single thread meandering channel, but with common gravel point-bars.

Two other Pyrennean rivers cross Languedoc plain (see Chap. 11, see also Calvet et al. Chap. 13 in Fort and André, Eds. 2014), the Têt (to the north) and the Tech (to the south). Both cross the Rousillon lowland, reaching the coast near and south of Perpignon. Within the Pyrenees both, especially the Têt, have important bedrock-gorge reaches (see Chap. 11). On emerging into the Rousillon basin at Ille-sur-Têt, where there are badlands known as "Les Orgues" (organ pipes!). From Ille-sur-Têt the river crosses the Rousillon basin to Perpignan and the coast at Canet. Throughout this reach the river has a low-sinuosity gravel-bed channel, locally with rudimentary braiding. There are two other rivers feeding into the Rousillon basin, both much smaller than the Têt. The Agly, to the north of the Têt, is fed largely from the folded Cretaceous rocks north of the Rousillon basin, within which it is incised within some impressive incised meanders. The Tech is in the south of the basin, fed from granitic terrain. It enters the basin at Le Boulou. It is a gravel-bed river with large low-amplitude meanders.

You may have noticed the abundance of "wine-related" local names in Languedoc. It is the most important wine region in France, not necessarily for great wines, but in terms of volume and medium quality wines!

Languedoc, the Coast   Virtually the whole of the Languedoc coast is a depositional coast, characterised by barrier beaches and lagoons (Fig. 14.9 map), with the exceptions of a few rock headlands. These rock headlands are at Sète (of Jurassic rock), at Agde (of Quaternary volcanic rocks), at the Montagne de la Clape near Narbonne (of Cretaceous rocks), and at Leucate (of Oligocene sedimentary rock). Finally, to the south of Rousillon, the coast is erosional in the eastern Pyrenean metamorphic and lower Palaeozoic rocks (see Chap. 11). On the coast the major lagoons (Fig. 14.9 map) are from north to south: the Étangs de Mauguio et de Vic (between la Grande-Motte, on the edge of the Camargue, and Sète); the Bassin de Thau (between

Sète and Agde); the Étangs de Bages et de Sigean and the Étang de l'Ayrolle (between Gruissan and Port-la-Nouvelle); and the Étangs de Lapalme and de Leucate or de Salses (between Port-la-Nouvelle and Perpignan). The main Languedoc rivers all enter the sea at very undistinguished river mouths, many across sand bars partially blocking the exits. They are from north to south (see above) the Rivers: Hérault (mouth near Agde); Orb (mouth at Valras-Plage); Aude (mouth south of Valras-Plage); and Têt (mouth at Canet-en-Rousillon).

Highlights of the Southern French Alps, Provence and Languedoc

The Ardèche Gorges (Fig. 14.2a, b) (see also Mocochain and Jaillet Chap. 19 in Fort and André, Eds. 2014) Cut deeply into the Cretaceous limestones set within the lower Rhône valley are the Gorges of the lower Ardèche downstream of Vallon-Pont-d'Arc. Vallon-Pont-d'Arc itself has a spectacular site (Fig. 14.2a), a natural arch related to a cut-off meander. The gorges themselves are tortuous incised meanders, some of the most spectacular in France (Fig. 14.2b). The presence of the deeply incised gorges has been related to significant base-level falls in the lower Rhône associated with the Messinian salinity crisis. I am not totally convinced by that argument, low Pleistocene glacial sea levels could have had the same affect. For much of the distance the modern channel of the Ardèche is a gravel flanked channel, but bedrock cliffs exhibit karstic features at a number of scales. There are excellent views through the length of the canyon from the D290 road above the north rim. Alternatively, at least during the summer, canoes can be hired from Vallon, with transport pre-arranged back from St Martin-d'Ardèche at the mouth of the canyon. Alternatively you can hike in for some distance along the canyon floor from Vallon, but I am not sure how far this is possible. A memory I have of the floor of the canyon is of naked sunbathers—in an area that seemed to be infested by poisonous snakes, vipers—bizarre!

Les Baronnies, Mont Ventoux and the Nesque Plateau (Figs. 14.3a, b, c) This area is located east of the lower Rhône valley, essentially from Dieulefit, south of the upper Drôme valley, southwards to the Plateau de Vaucluse and the Montagne du Luberon, and north of the lower Durance. It includes a whole range of spectacu-lar landscapes. The underlying geology is primarily of Jurassic "marnes noires" overlain by Cretaceous limestones. This rock sequence is folded in the north, by NW-SE aligned (Alpine trend) structures. In the south the rock sequence is folded by E-W aligned (Pyrenean trend) structures. The upper reaches of the Eygues (through Nyons), the Ouvèze (through Buis-les-Baronnies), and the upper reaches of the easterly-flowing Méouge (through Séderon), all have transverse reaches crossing fold structures in the Cretaceous limestones. Is their origin antecedence (from Palaeogene drainage)? Or is it superimposed from a Palaeogene unconformable cover stripped during the Neogene? Presumably their origin is similar to that of the northern Pyrenean drainage (see Chap. 11).

There is a hint of the Mediterranean in the upper reaches of each of these rivers with badland terrain cut into the "marnes noires" (especially near Ste. Jalle, Fig. 14.3a, see also Fig. 4.3a) and also with gravel-bed braided channels to the rivers. These rivers have well developed late Pleistocene river terraces within the main valleys.

Further south the east-west aligned mountain ranges are more pronounced: east of Vaison-la-Romaine, the Montagne de Bluye; south of Vaison-la-Romaine the dramatic "karstic" Dentelles de Montmirail (Fig. 14.3b); and above all Mont Ventoux (Fig. 14.3c). With a maximum elevation of 1909 m the summit areas of Mont Ventoux preserve Late Pleistocene periglacial patterned ground (see Fig. 4.3d). Further south is the extensive karstic Plateau de Vaucluse, dissected by the Gorges de la Nesque, south of which is the karstic spring, Fontaine-de-Vaucluse.

The Buëch Valley—Laragne/Sisteron Area (Figs. 14.4 Map, 14.5) The east-west structures continue east from the Baronnies area (see above) across northern central Provence in the Laragne/Sisteron area through the middle reaches of the Durance valley (Fig. 14.5a) and that of its tributary, the Buëch. The structure is very similar to that of the area described above, Upper Jurassic "marnes noires" overlain by Cretaceous limestone (Fig. 14.5b). The whole is folded into a series of east-west structures forming the breached anticline at Séderon, the synclinal fold at Serres north of Laragne; the mountain ridges that bound the synclinal Montagne de St Genis, and the syncline south of the Chabre ridge to the west of Laragne. At Sisteron an east-west thrust fault brings up Middle Jurassic limestones from the south forming a vertical wall breached by the south-flowing Durance River just downstream of the Buëch confluence (Figs. 14.4 map, 14.5a). Within the Laragne lowland near Upaix and Mison there are badlands cut into the "marnes noires". The same rocks crop out in the valleys between Veynes and Barcillonnette, north of Laragne. There are more badlands there. The Durance River is heavily manipulated for the generation of hydro-electric power, primarily through the Serre Ponçon dam and reservoir east of Gap. Although strictly outside the Laragne/Sisteron lowland, this reservoir is well worth a visit. Only completed in 1960, its upper reaches illustrate the high sediment yields from the Alpine Upper Durance. This has caused rapid delta growth into the upper part of the lake. There has also been substantial sediment input from the three main tributaries into the upper arm of the lake. From Serre Ponçon downstream the natural channel is still there on the valley floor, but it carries very little of the flow or sediment. The flow is canalised downvalley to feed a series of power plants.

The Buëch contrasts with the Durance in that its channel is almost entirely natural, wide shallow and mostly braided (see Fig. 4.5b). There is only one minor dam within the catchment, downstream of Serres, and that seems to trap very little sediment. The braids are best developed downstream from Laragne towards Sisteron. There is one other type of channel characteristic of this area, the incised meanders

of the Méouge, a western tributary of the Buëch incised into the Cretaceous limestone bedrock west of Ribiers (Fig. 14.5c).

The Valensole Plateau and the Verdon   Further south, three east-bank tributaries of
the Durance are interesting: the Bléone sourced in the crystalline high Alps east of
Digne; the Asse sourced south of Digne; and finally the Verdon sourced in the
Cretaceous limestone plateau near Castellane. Between each of the three are dissected plateaux on Cenozoic sedimentary rocks.

Lower reaches of both the Bléone and the Asse have (managed?) braided channels (Fig. 14.6b), but the Verdon has spectacular bedrock-incised reaches in three
main areas. The lowest downstream are the Basses Gorges, within which the river is
held up by a low dam, creating the Lac d'Esparron. Further upstream is the Barrage
de Sainte-Croix, holding up a large lake (Lac de Ste. Croix) which extends upstream
into the Grand Canyon of the Verdon. For most of the canyon the river is in a natural
partially bedrock-controlled course. There are access roads perched above the canyon on both sides, offering spectacular viewpoints, especially from the north side at
Pointe Sublime (Fig. 14.6c, see also Fig. 4.4b). Further upstream still, beyond
Castellane and extending to St André-les-Alpes, is the lake formed by the Barrage
de Castillon (Lac de Castillon). Above the lake the uppermost Verdon is a mountain
braided river draining a previously (Late Pleistocene) glaciated catchment within
the crystalline Alps.

The Mercantour Massif   Within the Alpes Maritimes for geomorphology avoid the
coast and go inland, up the valley of the Var towards the Mercantour massif. The
lower Var valley is a straight deeply incised valley, floored by a managed braided
channel. It runs counter to structure. After about 30 km upstream from the coast,
within the Défilé du Chaudan, the main channel of the Var flows in from the west as
a strike stream within the folded Cretaceous rocks. The Tinée flows into the Var
from the north as a transverse drainage, originating within the core area of the crystalline rocks of the Mercantour National Park. There are no modern glaciers within
the Mercantour, but there was Late Pleistocene cirque glaciation at the highest elevations. To see something of this landscape, take the Tinée valley to Isola (Fig.
14.7a). Note the tributary-junction alluvial-fan setting of Isola. There turn right off
the Tinée road towards the ugly ski resort of Isola 2000. At these elevations the
effects of Pleistocene glaciation are obvious: cirques, moraines, post-ice scree.
These contrast with the lower valley-side slopes which are deeply dissected by a
dendritic drainage network of steep low-order valley-side streams.

Valley of the Var   (Figs. 14.7b, c) If following on from the Mercantour itinerary
(above) return to the Défilé de Chaudan and the confluence of the Tinée with the Var.
Otherwise, head directly to that spot. Follow the N202 to the west which follows the
Var upstream along its strike-orientated course through Villars and Puget-Théniers
to Entrevaux (Fig. 14.7c) along a synclinal structure within the outcrop of Cretaceous
rocks, thrust from the south. Near Villars the Var has an incised bedrock channel, but
a little further upstream most of the narrow Var valley floor is occupied by the modern braided channel of the Var itself. There are a few localities where there are Late

Pleistocene low river terraces. At Entrevaux the Var is joined from the south by the extraordinary complex transverse drainage of the Chalvagne. This stream is sourced within a vale that wraps around the outer rim of a syncline in Cretaceous rocks. It then suddenly turns northwest to cross the scarp marking the lip of the syncline within a deeply incised valley towards the Var at Entrevaux.

The attractive village of Entrevaux marks a pronounced break in the valley geomorphology of the Var (Fig. 14.7c). Upstream of Entrevaux, towards Pont-de-Gueydan the river is transverse to an east-west fold structure in Lower Cretaceous limestone (Fig. 14.7b). To the west of Pont-de-Gueydan, in the Annot area, in the Var/Verdon divide zone, is another area of extremely complex transverse drainage. You could continue into this area by staying on the N202 to Annot and beyond, however a more exciting route (geomorphologically) would be to continue following the Var upstream.

If that is what you choose to do, turn right at Pont-de-Gueydan onto the D902 which follows the Var upstream. At that location (see Fig. 14.7b) the Var, which flows from the northeast from its headwaters west of the Mercantour massif, suddenly changes direction to flow towards the southeast. Is it possible that originally this whole area drained southwestwards towards the Verdon? The whole area is an area of remarkable transverse drainage. In addition to the previously described east-west folds in Cretaceous rocks, there is, to the northeast of Entrevaux, a domal structure bringing up gypsiferous (?) Triassic sandstones ad marls in its core (Fig. 14.7b). This structure is cut on its northwest flank by the transverse River Var within the Gorges de Daluis. Is the domal structure diapiric, a salt dome or a gypsum dome, destabilised by Neogene tectonics? If so, has that had an influence on drainage development? I do not know the answer to those questions. The modern River Var in these reaches has a gravel-bed braided pattern, interspersed locally by bedrock-gorge reaches, for example the Gorges de Daluis, upstream from Pont-de-Gueydan.

Beyond that gorge at Sauze Guillaumes you are spoilt for choice. You can either return by turning east along the D28 to Beuil then south through the Gorges du Cian to rejoin the N202 between Puget Théniers and Villars-sur-Var. Or from Beuil go east along the D30 to St Sauveur then south along the Tinée back to the N202. Each of these routes gives views of spectacular deeply dissected mountain country.

My preferred option from Sauze Guillaumes would be to follow the D2202 along the upper Var valley. Note the large (Pleistocene) tributary-junction alluvial fan at Villeneuve-d' Entraunes. Note also the extensive gravel-bed braided reaches. These are indicative of very high modern rates of sediment supply. This is confirmed beyond St Sébastien-d'Entraunes in the uppermost Var valley on the edge of the Mercantour National Park by the extensive gullied slopes cutting into (shaley) Cretaceous and Palaeogene rocks of the Pre-Alpine ranges on the margins of the Mercantour crystalline massif. Unfortunately, the only reasonable route out of this area is to return down the Var the way you have come. You could continue north over the Col de Cayolle towards Barcelonette. Alternatively, you could retreat to St Martin-d'Entraunes then head west via Annot and along the N202 towards the upper Verdon valley.

The Maures/Esterel Massifs As mentioned above, much of the coastal area of the "Côte d'Azur" is developed for resort or for private use rather than as a field laboratory for geomorphology. I suspect too that not many people come here for the specific purpose of geomorphic field work. Having said that, there are areas of the Maures and Esterel Hercynian granitic masses that are less developed as resort or private areas. They exhibit rather low-key, dominantly erosional coastal geomorphology (Fig. 14.8a), locally preserving late-Pleistocene raised wave-cut platforms.

Three areas of easy access to the coast exist, two on the Maures massif and one on the Esterel massif, each area characterised by low bedrock cliffs and small bayhead beaches. These areas are on the Maures between Le Lavandou and La Croix-Valmer and between Ste. Maxine and St Aygulf, and on the Esterel between Agay and La Napoule.

The Eastern Camargue (see Arnaud-Fassett and Provensal, Chap. 20 in Fort and André, Eds. 2014) To see more of the diversity of the Camargue it is best to drive the back roads, visiting a series of sites, rather than focussing on individual specific sites. I suggest two itineraries, one through the eastern Camargue, the other through the centre and west of the Camargue. For the eastern Camargue if you start at Avignon, an interesting city in its own right, site of the famous "Pont d'Avignon", head south to Arles. This is where the Camargue really starts (Fig. 14.10 map). En route you can either go direct (via the N570 route) or follow the Rhône on a variety of back roads. Just north of Arles is where the diminutive Petit Rhône distributary leaves the main river heading towards the western Camargue (see below).

On this trip, from Arles make sure you cross the river to the west bank then head south on the D570. After about 1 km turn left onto the D36 towards Salin-de-Giraud continuing along the west bank of the Rhône. After a couple of km take the right hand fork along the minor road, the D36a, towards Villeneuve and Pont Noir. As far as Villeneuve much of the Camargue wetland has been reclaimed for agriculture (especially for rice), but from there on it is in at least a semi-natural state. At Pont Noir, on the edge of the enormous deltaic lagoon, the Etang de Vaccarès, there is an ecology information centre. Indeed, one of the attractions of the Camargue as a whole is its wildlife, particularly its bird life. From here, continue south on the D36a/D36c towards Salin-de-Giraud. After leaving the shores of the Etang de Vaccarès the terrain becomes increasingly deltaic, with the road following wetlands that mark former meandering distributary channels of the Rhône. Salin itself is disappointing. It is a centre of salt production. It is almost on the delta coast, but when I was there several years ago it was impossible to get anywhere near the sea.

The Central and Western Camargue Starting from Arles, follow the D570 on the west side of the Rhône towards Albaron and Stes Maries-de-la-Mer (Fig. 14.10 map). Continue on this road across the reclaimed northern part of the Camargue to Albaron which is on the main distributary channel in the delta, the meandering channel of the diminutive Petit Rhône. The road then follows this meandering channel for several kilometres before turning south through lagoon country (Figs. 14.9 map, 14.10 map, 14.11) and wetlands to Stes Maries. There are two routes from the road junction at Pioche Badet, the eastern route follows the D85a through the wetlands by the Etang de l'Impérial. Alternatively stay on the main D570 via the "Parc Ornithologique" and information centre direct to Stes Maries. Stes Maries is a busy coastal resort, with beaches and marinas. It has Van Gogh connections. It is busy because it is one of the few places in the Camargue where a metalled road reaches the coa st.

Leaving Stes Maries to the west take the D38 which follows the coast for 1 km then turns inland following the Petit Rhône River to the Pont de Sylvéréal. There turn left (west) over the bridge towards Aigues-Mortes, more or less on the western edge of the Camargue. It is a small town on the Sète/Rhône canal, 6 km inland from the coast. You could continue then 6 km southwest to Le Grau-du-Roi on the coast. This is a heavily developed part of the coast for tourism with beaches and marinas between Port-Camargue and La Grande-Motte. However, it does have some under-lying coastal depositional geomorphology, sand spits dunes etc., especially east of Port-Camargue. From this area it is best to head towards Montpellier following the coast road between the beaches and the Etang de l'Or.

The Languedoc Coastal Area, South from Montpellier *The Northern Languedoc Lagoon Coast South from Montpellier to Agde* This is a lagoonal coast. Inland it is an area of vine cultivation. Further inland are the low mountains of Haut Languedoc. South of Perpignan is the very eastern part of the Pyrenees. I do not know this area well, having driven through only a few times, so I can recommend only a few specific sites or areas.

From Montpellier to Sète the coast is marked by almost continuous lagoons (Fig. 14.9 map), particularly the Etang de l'Arnel and the Etang de Vic. A ridge of Jurassic limestone separates these lagoons from the Languedoc coastal plain inland. I do not think it is possible to drive along the coastal barrier from Palavas-les-Flots to Frontignan and Sète, but access to the shore (sandy beaches) is possible at Villeneuve/Ile de Maguelone and from Vic-la-Gardiole. Sète is an important port, beyond which there is another large lagoon, the Bassin de Thau behind a coastal barrier, before you reach the Cap d'Agde (I've never been here but it sounds like a bizarre place) and Agde itself. The geology here sounds interesting, a boss of Quaternary volcanics. There is also variety in coastal morphology, with the modern mouth of the now canalised River Hérault to the west of the town and its former course into salt-marshes east of the town.

The Southern Languedoc and Rousillon Coast from Agde to Argelès-Sur-Mer South from Agde, there is a break in the lagoon coast. Instead the coast is of low sandy beaches, broken by occasional inlets, until the mouth of the Orb at Valras-Plage. Immediately beyond there the mouth of the Aude and its associated marsh-lands prevent through access. It is only beyond St Pierre-sur-Mer/Narbonne-Plage that continuous access along the coast is again possible. Even then access only lasts as far as Gruissan, essentially along the coastal strip, which is seawards of the Cretaceous limestone ridge of the Montagne de la Clape.

South of Narbonne and Gruissan are more lagoons, the Étang de Barges et de Sigean, the Etang de Lapalme and the Etang de Leucate/ Salses, all between Narbonne and Perpignan. Coastal access is possible at Port-la-Nouvelle, then at

Leucate, from where there is continuous coastal access throughout the Rousillon lowland until Argelès-sur-Mer where the eastern Pyrenees meet the coast. Access is unimpeded, even by the (insignificant) Pyrenean-river outlets of the Rousillon: the Agly, the Têt and the Tech.

Southern Languedoc and Rousillon: River Systems   Although I hardly know this terrain, I am sure that the main river valleys of Languedoc/Rousillon would be worth exploring. These are from north to south as follows.

The Hérault This river drains the Causses area of the southern Massif Central (see Chap. 8), skirting the Cévennes and crossing the Languedoc plain to reach the sea at Agde.

The Orb This river rises in the mountains of Haut Languedoc and crosses the Languedoc Plain through Béziers to reach the sea near Valras-Plage.

The Aude This river drains the Carcasonne gate area, being fed particularly from the southern (Pyrenean) side of the valley and receives tributaries also from that side, such as the Orbieu. The Aude flows north of Narbonne reaching the coast just south of Valras-Plage.

The Agly, the Têt and the Tech (see Chap. 11).

These three rivers drain the eastern Pyrenees, then cross the Rousillon lowland basin, reaching the coast east of Perpignan.

Chapter 15
Corsica

Although an offshore island, Corsica is geologically and geomorphologically, as well as politically, an integral part of France (unlike so-called "France d'outre-mer", most of which is/are tropical islands, which demand their place, quite irrelevantly in any compendium work on the geomorphology or geology of France—a bee in my bonnet!).

Geomorphologically, Corsica must rank as one of the most exciting regions of France. Not only does it have marked structural and geological contrasts between the western (former Hercynian) and eastern (Alpine) parts of the island, but touristically it is much less well developed than most parts of mainland France (or was when I visited it many years ago). For example the spectacular western coastal areas are (or were) almost undeveloped. A word of warning: apart from a few main roads between the major towns, the road system is (or was) primitive, characterised by single track narrow roads with numerous hairpin bends. Drive with care! Corsica is also costly and awkward to reach: fly and hire a car, or use a long (and expensive) sea crossing by car ferry from one of the southern French mainland ports (Nice or Marseilles). To see the landscape you need car transport. There are a few bus services, and an interesting, but slow and limited rail network (I'm not sure if that is still operational). I only know well the northern part of the island, so most of the material presented here relates to that area, with only a basic cover of the south.

The Geology of Corsica (Fig. 15.1 Map)

Corsica is part of Alpine Europe, but has two component parts (Figs. 15.1 map, 15.2a, b). The west is an enormous piece of metamorphosed granite and other rocks metamorphosed by Hercynian mountain-building during the late Devonian and Carboniferous periods. Then much later, during the Eocene, Western Corsica was incorporated into Alpine Mediterranean structures. Western Corsica, with

© The Author(s), under exclusive license to Springer Nature
Switzerland AG 2025
A. Harvey, *The Geomorphology of French Landscapes*,
https://doi.org/10.1007/978-3-031-68490-6_15

neighbouring Sardinia, was originally part of the Iberian massif, which rotated and became thrust towards the east (Fig. 2.4 map). This was partially as a result of the eastward collisional movement of the African plate against the southern margin of the European plate during the end-Cretaceous to Palaeogene (more or less contemporaneously with the formation of the Pyrenees). This process incidentally opened up the western Mediterranean marine basin.

Eastern Corsica comprises mostly Jurassic and Cretaceous low-grade metasedimentary rocks together with oceanic crust igneous rocks, all deformed by Alpine ('Mid-Tertiary') structures. This unit is related to the Italian Appenines and was thrust in from the east against western Corsica, during the Miocene. It was then subject to further rift-like extensional faulting creating a complex N-S fault belt through the centre of the island (Figs. 15.1 map, 15.3a, b).

Virtually the whole of western Corsica comprises beautiful metamorphosed granitic rock, characterised by spectacular granite scenery (Figs. 15.2a, 15.4a, b). In the northwest (south of Calvi, north of the peak of Monte Cinto 2706 m) is a volcanic structure of late Carboniferous to Permian (ie. late Hercynian) age, incorporated within the granite terrain. This structure is visible in the rocks rather than being obvious in the modern terrain, although the former lip of the caldera does form a modern mountain ridge (Fig. 15.2a), and apparently formed the backwall of a cirque during the penultimate Pleistocene glaciation. Apart from this structure the main structures affecting the west are a series of WSW-ENE aligned shear faults, truncated in the centre of the island by the (Miocene?) central Corsican fault zone (see below and Fig. 15.3a, b).

The central Corsican fault zone is a major feature in the geology of Corsica (see Figs. 15.3a, b). The fault zone is aligned more or less north-south, and separates the Hercynian granitic rocks to the west (Figs. 15.2a, 15.4a, b) from the mostly Jurassic metasedimentary and igneous Alpine rocks to the east of the fault zone (Fig. 15.2b). Within the fault zone itself the earlier structures are relatively low-angle E-W thrusts which were later truncated by normal (extensional) faults creating a linear N-S aligned tectonic depression (an ill-defined rift) separating the two terrains (Figs. 15.1 map, 15.3a, b).

Post-Alpine rocks include a small area at Bonifacio on the southern tip of the island of near- horizontal, post-Alpine (Late Miocene), thinly bedded limestones which rest unconformably on the Hercynian granitic rock. They form a small patch of lowland bounded by low coastal cliffs which are dominated by spectacular topple failures. The eastern coastal plain near Aléria is also formed of upper Miocene sedimentary rocks, locally overlain by Pliocene sediments. The small area of coastal plain further north (south of Bastia) is entirely Quaternary in age.

The Geomorphology of Corsica

(Figs. 15.1 map, 15.2a, b, 15.4a, b, see also Le Coeur, Chap. 23 in Fort and André, Eds. 2014). Corsica is a mountainous island, with a maximum elevation of 2706 m at Monte Cinto. On that mountain there is evidence of localised small-scale

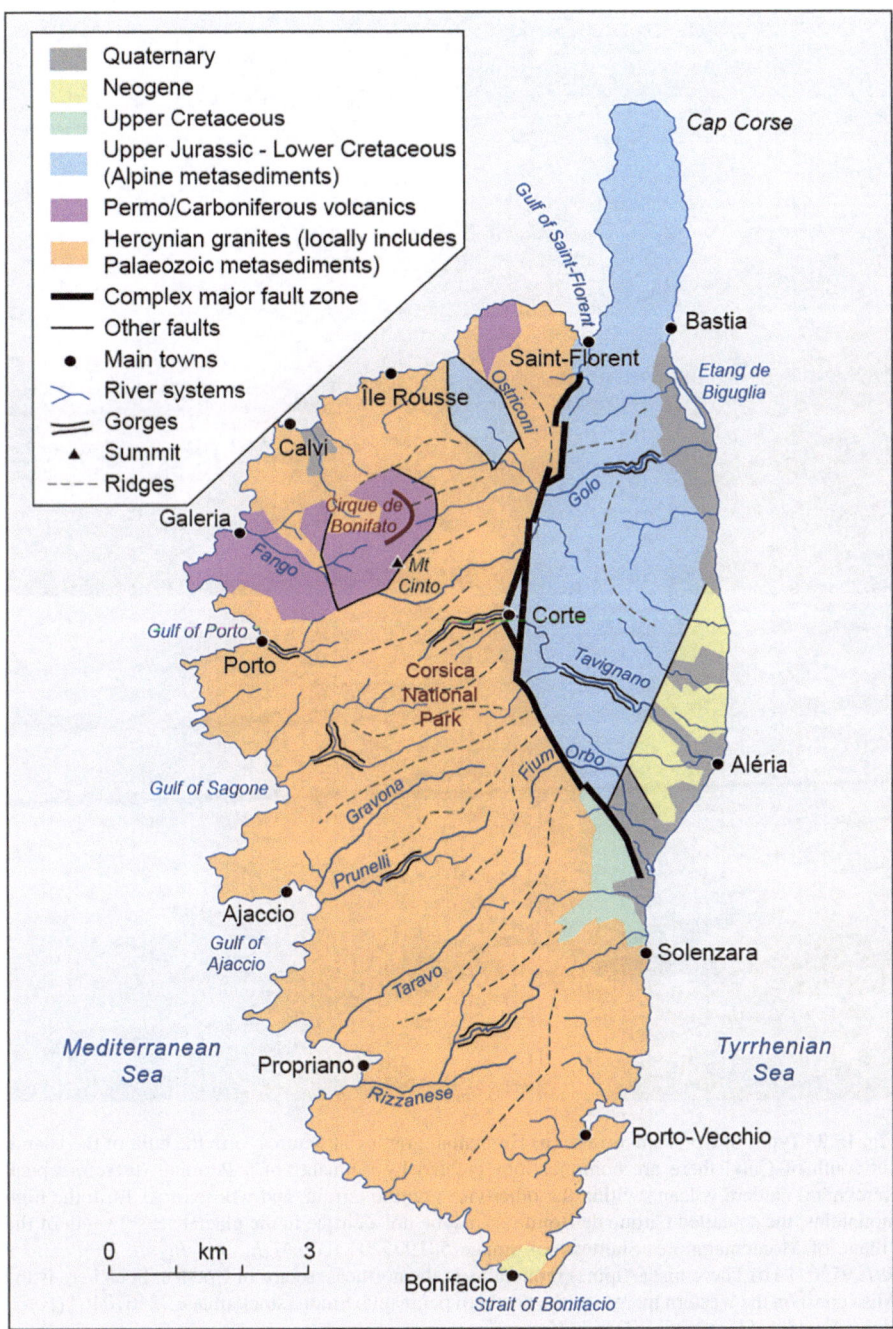

Fig. 15.1 Map of Corsica

Fig. 15.2 Types of Corsican terrain. (**a**) Hercynian granitic structures form the bulk of the island, but south of Calvi there are complications created by remnants of a Permian (therefore peak Hercynian) caldera volcano within the otherwise granitic terrain, and whose rocks form the high mountains, the so-called Cirque de Bonifato (maybe not a cirque in the glacial sense) south of the village of Montemaggiore: Shutterstock image 581995204 (*copyright: Jon Ingall, Shutterstock 581995204*). (**b**) The schist (Alpine) mountains in the northeast sector of Corsica. Seen here is the Alisa coast on the western margin of the northern peninsula: Shutterstock image 724670464 (*copyright: Naeblys, Shutterstock 724670464*)

Fig. 15.3 Corté, the ancient capital of Corsica, sitting astride the central Corsican fault zone. (**a**) Landsat image of the Corté area [E 9.09' N 42.18']. The fault zone is aligned NE-SW to the SE of the city of Corté. Then it is aligned NS through the city and to the north. Note the contrast between the rugged Hercynian granite terrain to the west and south of the fault zone and the much smoother terrain on the Jurassic and Cretaceous metasediments and sedimentary rocks to the east and northeast. (**b**) The medieval citadel in Corté is perched on a linear ridge above the city. This ridge appears to be defined probably by two faults within the central Corsican fault zone: Shutterstock image 1059439517 *(copyright: Olezzo, Shutterstock 1059439517)*

Pleistocene glaciation in the form of a few small degraded cirques (eg. Cirque de l'Asco), small glacially-scoured lake basins, and some small-scale (not very distinct) morainic features. There also appear to be glacial features (cirques and lake basins) in the next main range to the south, the Monte Rotondo (maximum elevation 2622 m), and further south in the Gravona ranges (Conchon 1978, 1986; Kuhlemann et al. 2005). It is probable that these features are of penultimate glacial age, but there are suggestions of last glacial features of more limited extent. Otherwise, these mountain areas are dissected by the stream system, within the 'Hercynian' western

Fig. 15.4 Geomorphic features of the granite mountains of western Corsica (see also Fig. 4.1b).
(**a**) Pressure-release joints near Calenzana, south of Calvi: a mechanical weathering process related
to the late Neogene to Quaternary erosional and incisional offloading of massive granite. This
process resulted in the formation of slope-parallel valley-side granite slabs. (**b**) The Bonifato area
is drained by the N-flowing Figurella river system. Beyond the mountains it is characterised by a
wide shallow cobble-bed braided channel. The bed sediment comprises beautiful rounded granitic
and volcanic cobbles

part of the island often in deep rugged bedrock gorges in a general NE-SW alignment with faults.

The terrain in the eastern (Alpine) part of the island is more subdued, mostly
scrub-forest covered, and dissected by deeply incised valleys, many with main
channels in incised meanders. Within the headwater catchments in this part of the

island, there is mostly a clear dendritic drainage network of small tributary valleys feeding the main drainages. There is relatively little rock exposure in the headwater catchments in this part of the island, whereas on the granitic terrain to the west, bedrock exposure characterises the main ridges and the rugged slopes of the head-water basins.

Below I give a general description of the geomorphology of the island, mountains to coast, working anti-clockwise from St Florent on the Golfe de St Florent at the southwest corner of the northern peninsula (about 18 km west of Bastia). St Florent is just east of the fault zone separating the Mesozoic (Alpine) rocks from the (Hercynian) granites to their west (Fig. 15.1 map). To the west of the fault zone in this area is a zone of low granitic hills (just over 400 m high), the Désert des Agriates (not a desert, simply empty country). Its almost inaccessible rocky coastline is one of low cliffs with occasional small bayhead beaches, and inlets fronted by shingle spits.

To the west of the Désert des Agriates is a small infaulted area of "Alpine" Cretaceous rocks, with characteristic shallowly dissected terrain. The faulted boundary is picked out by the SE-NW aligned drainage of the Ostriconi. Beyond (ie. further to the west of the "Désert des Agriates") lies the Balagne lowland between L'Île-Rouse and Calvi. 'Lowland' is a relative term—the area reaches elevations of over 500 m. However these are in contrast to elevations of over 2000 m in the mountains behind. These mountains are formed of Late Hercynian volcanics (see above and Fig. 15.2a). The southern margins of these volcanic rocks culminate in the Monte Cinto range, the highest mountains in the island (reaching a height of 2706 m). This is one of the three mountain ranges on which remnants of Pleistocene glacial features have been reported (see above). The Balagne is developed in beautiful Hercynian meta-granites, which as elsewhere in the granitic terrain, exhibit a range of weathering features (see Chap. 4) including: pressure-release jointing (Fig. 15.4a) which resulted in slope-parallel thick slabs of granite. These were subjected to long periods of (Neogene or early Quaternary) deep weathering (see Fig. 4.3e); then case-hardening and cavernous (honeycomb) weathering (see Fig. 4.1b). There is also evidence of (later) Pleistocene hillslope solifluction (see Fig. 4.3e), and much later human modification of the detail of some steeper hillslopes in the form of agricultural terraces.

The steep bedrock mountain slopes feed rounded boulders, cobbles and pebbles into the streams (Fig. 15.4b) and ultimately onto the shingle beaches of the area. The main drainage of northwest Corsica is the Figarella, which together with the neighbouring Calanzana drainage, reaches the sea in the Golfe de Calvi, just east of Calvi (Fig. 15.1 map). Both drainages rise within the mountains. In the upper reaches, particularly of the Figarella, they are characterised by incised bedrock channels. Where the rivers enter the Calvi lowland a broad terraced alluvial fan-like depositional form is created, across which the modern channels have cobble-bed braided patterns (Fig. 15.4b). Palaeochannels, the oldest of which are presumably early Pleistocene in age, can be traced on the older fan surfaces.

The coastal geomorphology differs between north coast and the higher energy west coast. East of Calvi, the north coast is dominated by low granite cliffs, locally preserving former rock platforms (see Fig. 3.7d) and raised beaches. Their age of these features is uncertain. They presumably relate to a Pleistocene high sea level (perhaps during the last interglacial, 100 ka BP), or in the case of at least some of the raised beaches, to a mid-Holocene post-glacial maximum sea level (7 ka BP)? Locally, both east of L'Île Rousse and at Calvi there are modern depositional zones, with minor spit and bar forms.

West of Calvi the coast is an irregular granite cliffed coast. The whole of the west coast south from Calvi to Ajaccio is spectacular, actively eroding granite cliffs with occasional bayhead beaches (Fig. 15.5a). South of the late Hercynian volcanics (ie. south of the Golfe de Porto) the coastal cliffs are formed of a variety of Hercynian granitoid and granodiorite plutonic rocks (Fig. 15.5b), which locally have weathered into spectacular pinnacle forms. Between headlands there are a few small (mostly undeveloped) bayhead beaches or river mouths (Fig. 15.5c). There are occasional raised Pleistocene shoreline features.

Inland, south of Porto, are a series of faults, orientated NE-SW, which have to a large extent been picked out by the main drainages (Fig. 15.1 map). These drainages are from north to south: the Porto/Tavulella system (which drains the south side of the Monte Cinto range); the Fuime Grossu and Liamone/Cruzini systems (which drain the south side of the Monte Rotondo range); and the Gravona system (which drains the south side of the Monte d'Oro south of Monte Rotondo, and reaches the coast at Ajaccio). In their headwaters these drainages tend to have bedrock channels incised into deep valleys. In their middle reaches they locally have steep incised bedrock reaches, or locally braided channels within incised valleys. They enter the sea across small depositional lowlands.

South of Ajaccio (Fig. 15.1 map) the inland relief on the granitoid bedrock is less than that north of Ajaccio, with elevations between Ajaccio and Propriano reaching only 629 m. Elevations are even less in the coastal area between Propriano and Bonifacio on the south coast and Porto-Vecchio on the southeast coast. South of Ajaccio the terrain is still dominated by NE-SW orientated granitoid ridges partly separated by similarly orientated fault systems. There are two main NE-SW orientated drainages in this area, the Tavaro in the north and the Rizzanèse in the south. Both reach the sea in the Golfe de Valinco, near Propriano. Both are sourced in the high granitic terrain in the central part of the island. The Tavaro rises east of, and the Rizzanèse rises south of, Monte Incudine (2134 m). Both rivers have incised upper and middle courses within bedrock channels. Only lower downstream is there any valley floor and any modern gravel accumulation. In the most seaward reaches, of the Tavaro especially, there is a Late Pleistocene to Holocene low angle valley-fill/ fan-delta through which the modern channel meanders and across which there are palaeochannels evident on the surface. On the Rizzanèse the upstream incised reaches give way to an alluvial valley floor near Ste. Lucie-de-Tallano, further upstream than on the Tavaro. Within this alluvial reach the channel is sinuous, but with plenty of local gravel accumulation. The coastal fan-delta zone is less extensive and less clear than that on the Tavaro.

The southernmost part of the island, south of a line roughly between Sartène and Porto Vecchio, is generally low lying. It comprises a series of low NE-SW

Fig. 15.5 NW Corsica: coastal forms (see also Fig. 3.7d: coastal rock platform near Calvi). (**a**) Corsica west coast: This higher energy coast is a combination of cliffs cut in granite and lowland bayhead beaches or river mouths. (**b**) Granite cliffs: headlands on the west coast are cliffs in granites showing very irregular morphology, where weathering and erosion have exploited the weaknesses in the rock. (**c**) Bayhead beaches: These are depositional zones, here at Crovani Bay south of Calvi a shingle beach completely isolates a lagoon from the open sea

orientated granite ridges, culminating in the low cliffs of the south coast. In the far south, at Bonifacio, late Miocene horizontally bedded limestones rest unconformably on the Hercynian granitic rocks, forming spectular cliffs. For archaeology enthusiasts the whole of this southern lowland area includes numerous Bronze-age (?) menhirs, dolmens and stone alignments.

In the south of the island the **ea**st coastal area, north from Porto Vecchio to Ghisonaccia comprises low granite cliffs with numerous small bayhead beaches. The coast is backed inland by granitic mountain terrain, drained by short incised NE-flowing streams. This pattern changes north of Ghisonaccia. Between there and Corté (Fig. 15.3a, b) in the centre of the island, the Alpine-front fault system lies between the Hercynian granitic terrain in the west and the folded Alpine rocks in the east. The fault zone is aligned NW-SE (Fig. 15.1 map). The Alpine terrain is also fault-bounded to the east by a NNE-SSW orientated fault. Seawards of that fault is a lowland on a wedge of (post-tectonic) upper Miocene sedimentary rocks, locally overlain by Pliocene sediments. The coast itself is a depositional coast with spits and lagoons. This coastal lowland receives two major drainages. In the south is the Orbo, which rises in the Hercynian granitic terrain in the Monte Renoso area to the southwest of the Alpine fault zone. The Orbo then follows this zone to reach the east coast near Ghisonaccia. In the mountains it has a characteristic incised bedrock channel, but on emerging from the mountains has a wide gravel-bed meandering channel, locally set below Late Quaternary terraces.

Much more important, and in fact the largest drainage on the island, is the Tavignano. Its main headstreams rise within the Hercynian granitic area west of Corté, to the north of the Monte Rotondo massif. The headstreams converge in the Corté area. The main stream then flows south, more or less parallel to, then crossing the Hercynian/Alpine fault zone. This river has a narrow floodplain within incised meanders. Near Pont Génois it picks up a major west-bank tributary, the Vecchio. This river rises on the southern flanks of Monte Rotondo, then crosses the fault zone. Reach for reach, its channel is similar to that of the Tavignano. From Pont Génois downstream the Tavignano has a bedrock channel within gorges incised across the grain of the Alpine structures. Further downstream the gorge takes the form of incised meanders floored by a gravel-bed channel. Between Fajo and Casabertola it crosses the faulted Alpine mountain front onto the coastal plain, at which point the Tavignano gains an important tributary, the Corsigliese. In the mountain-front zone both streams have wide gravel-bed channels, hovering between meandering and braided habits, set below distinct Late Quaternary terraces. Further downstream irregular meandering dominates. The Tavignano reaches the sea near Aléria.

Further north the coastal plain is crossed by a series of relatively small drainages emanating from the Alpine highlands (Fig. 15.1). On reaching the plain, the valley floors are set below Late Quaternary terraces. There is also evidence of formerly active tributary gully systems, both as hillslope dendritic gully systems and as valley-bottom linear gullies.

At Ste Lucie de Moriani the plain narrows and the Alpine rocks almost reach the coast. North of here towards Bastia is another (more recent) coastal lowland, formed of Late Quaternary sediments and fronted by a spit and lagoon coast. That lowland receives two main drainages. The Alto in the south is a small drainage emanating from within the Alpine terrain. Within the mountains it has a characteristic incised mostly gravel-bed channel. It meanders in the short reach across the coastal plain with evidence of former meanders. In the north the Golo is a much larger river. Like the Tavignano it rises on the Hercynian granitic terrain to the south of Monte Cinto. It then flows north, parallel with the fault zone to Ponte Leccia. It then flows east across the Alpine terrain. In the headwater granitic terrain it has a characteristic mountain channel, locally cut into bedrock, otherwise locally with a gravel bed. Near Ponte Leccia it picks up several tributaries. Upstream of Ponte Leccia it collects the Casaluna from the east from the Alpine area, a stream with mostly an irregular incised meandering channel. In Ponte Leccia itself the Golo is joined by a system from the west from the granitic terrain, the Navaccia/Tartagine system, with a mixture of channel types including headwater bedrock reaches, mid-valley braids (on the Tartagine) and incised meanders (on the Navaccia). From Ponte Leccia, across the Alpine terrain, the Golo has irregular incised meanders locally floored by bedrock channels, and elsewhere by wide (in some places braided) gravel-bed channels. Lower downstream the valley is wider with evidence of abandoned cut-off channels, and with the modern channels set below Late Quaternary terraces. At the mountain front, at Casamozza, the channel is trenched into the surface of the coastal plain (in the style of a trenched alluvial fan) with terraces within the trench. The modern channel is braided, but further downstream towards the coast, entrenchment decreases, so that the channel is on the surface of the coastal plain, and the braided pattern gives way to meanders.

North of Bastia (Fig. 15.1), the northern peninsula is formed in part of Alpine metasedimentary and (basic) igneous rocks. Its east coast is mostly of low cliffs below steep coastal slopes. There are a few bayhead beaches at the mouths of the short W-E valleys. In contrast, the west coast is mostly a cliffed coast with some high cliffs, and apart from a couple of bayhead beaches in the southwest of the peninsula, it is a high-energy rock coast. Inland the peninsula is mountainous with several peaks rising to elevations of between 1000 m and 1305 m, and a number of others with elevations over 600 m.

Highlights of Corsica

NW Corsica, the Calvi Area (see also Le Coeur, Chap. 23 in Fort and André, Eds. 2014): The Balagne "lowland" (only a relative term!) and the Figarella valley from the Bonifato Mountains to the coast at Calvi, offer a wide range of geomorphologically interesting sites. At one end of the scale are weathering features, especially on the Hercynian granites of the Balagne and adjacent mountain slopes. These range from massive pressure-release joints (Fig. 15.4a, see also Chap. 4) to deep exposures of rotted granite (Fig. 4.3e) relating to warm humid conditions in the Late

Neogene or Early Pleistocene. In places these sediments are overlain by Late Pleistocene periglacial (solifluction) deposits (Fig. 4.3e). Locally, exposed rock faces exhibit cavernous weathering below surface case-hardening (Fig. 4.1b).

The Figarella rises in the high mountains of Bonifato (Fig. 15.2a: Carboniferous-Permian volcanic rocks within the Hercynian granitic area). The volcanics and the granites provide boulders which wear to beautiful rounded cobbles (Fig. 15.4b) within the braided channel of the Figarella. On emerging from the mountains the Figarella crosses a large late Quaternary terraced fan-like feature, within which lies its modern cobble-bed braided channel.

The coastal geomorphology in this area is also varied, ranging from depositional morphology within the Golfe de Calvi to an erosional coast of low granite cliffs near L'Île Rousse. Also on this section of coast there are localities with Pleistocene raised rock platforms (see Fig. 3.7d), some locally capped by raised beach sediments, probably relating to high sea levels during the last interglacial. West of Calvi there are impressive, high, granite cliffs (Fig. 15.5b), interspersed as at Crovani by bay-head shingle beaches (Fig. 15.5c).

Central Corsica, Corté and the Golo Valley I do not know this area well, but from what I have seen, it is an interesting impressive area full of contrasts. The first contrast is between eastern and western structural parts of the island. Corté (Fig. 15.3a), the ancient capital of Corsica, sits on the fault zone that separates the two (Figs. 15.1 map, 15.3b). To the west are the Hercynian granites, to the east are the folded Alpine (dominantly) metasediments. I do not know whether the faults themselves can be clearly seen; the locations are evident, and there is a clear contrast in the terrain on either side. To the west the granite terrain is rugged, with a lot of bedrock exposure. Elevations within 10 km of Corte reach almost 2000 m. The two incised valleys to the west of Corte are the main headstreams of the Tavignano, the upper Tavignano itself, and the Restonica. Both valleys are SW-NE fault-aligned. Both channels are dominantly bedrock incised. Beyond Corte the Tavignano drains towards the SE through the more subdued "Alpine" terrain of the eastern part of the island. Within 10 km to the east of Corté the relief barely reaches elevations of 1000 m. That terrain has a general north-south structural alignment (Fig. 15.1 map) with little bedrock exposure. The tributary headwaters have a relatively high drainage density in a broadly dendritic pattern. The Tavignano itself has a mixture of bedrock and gravel-bed channel reaches. Further downstream it has an incised meandering valley within which there are Pleistocene terrace remnants on the insides of the valley bends. The channel is mostly gravel-bed.

If you head north from Corté, either by the main road or the D18 side road, you soon cross the watershed into the Golo drainage. The Golo drains the area northwest of Corté, in rugged granite terrain on the flanks of Monte Cinto (2706 m), the highest point on the island. Its channel is mostly an incised bedrock channel. From Francardo downstream to Ponte Leccia it more or less follows the alignment of the fault zone that separates the western granitic part of the island from the eastern "Alpine" part. From Ponte Leccia the Golo turns east across the "Alpine"sector, crossing the north-south structural grain of the country. Its channel is similar to that of the Tavignano, at first a mainly bedrock-incised channel. Further downstream the valley has incised meanders, with Pleistocene terrace remnants on the insides of the bends, and a modern mostly gravel-bed channel.

Southern Corsica, Ajaccio to Bonifacio I do not know this part of the island, but from all reports it is well worth a visit. The coastal geomorphology is interesting; granite cliffs on headlands, sandy bayhead beaches, especially between Ajaccio and Propriano (Fig. 15.1 map). If you are also interested in archaeology, south of Sartène there are several important megalithic sites. There is also a Museum of Prehistory in Sartène itself. Bonifacio, on the southern tip of the island has a spectacular site, set on and against impressive layered Miocene limestone cliffs.

Chapter 16
Final Thoughts

One of the points I made in the Preface to this book is that landscape is not simply a response to geology (rock type), as is often simplistically argued in guide books, even in some text books. Landscape is the product of three phases which may be separated from each other by long periods of geological time. The three phases are: (1) the formation of the rocks themselves; (2) their structural deformation and uplift; and (3) their erosion to create the landscape we observe. It is true that both rock type, and tectonic deformation and uplift may be <u>expressed</u> in the landscape at both gross and detailed scales, but the timescale of the erosional landform development itself is at most only a few million years, rather than the several hundred million years of even only the post-Precambrian part of the geological timescale.

In the French context, from a broad geological point of view rather than from a narrower geomorphological one, the salient events were as follows (Chaps. 2 and 3, see also Fig. 2.1).

(i)   The Late Carboniferous to Permian plate-tectonic collision created the Hercynian mountain system, remnants of which, after eons involving uplift then erosion, now form the structures within the uplands of Armorica, the Ardennes, the Vosges and the Massif Central.

(ii) The Late Cretaceous to the Earliest Palaeogene saw the beginnings of the modern plate tectonic context with the opening up of the Atlantic Ocean. For France, the main effects were the rotation of Iberia to initiate the forming of the Pyrenees. There was also widespread deformation and uplift throughout France.

(iii) In "Mid-Tertiary" (primarily within the Miocene) the <u>present</u> plate-tectonic patterns became established. The Alpine system and the Mediterranean were created in something like their present form. Substantial uplift and some deformation took place throughout France.

All of that is Geology! It is only after the mid-Miocene that erosion produced the present landscape. In other words, it is only since then that the modern geomorphology developed (Fig. 2.4). The significant periods in the development of the modern landscape of France were as follows (see Chap. 3, see also Fig. 2.5).

(i) During the late Neogene away from the specific areas that were clearly geologically and therefore geomorphically active, most of France was geomorphically fairly stable. The active areas included areas of tectonic activity (particularly the Alps and the Pyrenees), and areas of active volcanicity (Auvergne). Elsewhere, extensive, low-relief erosion surfaces developed. These were formed particularly across the Palaeozoic rocks of the ancient Hercynian massifs, but also across the Mesozoic rocks (mostly limestones) of the intervening areas (essentially the Paris and Aquitaine basins). Mature soils on these surfaces indicate long periods of stability under (probably) warm, humid climatic conditions (Fig. 3.3d).

(ii) During (probably) the Late Pliocene or Early Pleistocene these planation processes (whatever they were) ceased, and almost all rivers began to incise. This may have been a response to Neogene post-tectonic uplift, but the direct cause was probably climatic deterioration. Also low base-levels related to low sea levels during the ensuing Pleistocene glaciations may have had an effect in coastal areas. Continued dissection was to dominate the Pleistocene.

(iii) The Mid and Late Pleistocene, globally, were dominated by a sequence of alternating glacial and interglacial conditions. We do not know the effects of the earlier glaciations on France, only of the last two major glaciations, the most extensive peaking at c150 ka BP and the last, peaking at about 20 ka BP. On both occasions there was an extensive ice cap over the Alps and Jura, a smaller ice cap over the Pyrenees and smaller glaciers in Auvergne, the Vosges and Corsica. Outside the glacial limits permafrost was extensive, affecting slope processes. River systems were also very active carrying high sediment loads (see Figs. 3.6 and 3.7).

(iv) In the early Holocene from about 10 ka BP, following the last glaciation, climate ameliorated to something like the present climate, albeit with short-term fluctuations. Sea levels rose to something like present sea level (by about 7 ka BP). Inland, woodland vegetation colonised most previously bare areas. Erosion rates and sediment supply to the river systems markedly decreased and the rivers effectively "shrank". The landscape stabilised. By the Mid to Late Holocene, things had begun to change. Humans had begun to use the land, especially for agriculture, radically affecting erosion rates. There are links between Holocene geomorphology and archaeology. As history progressed the human impact was ever increasing towards the present day. One wonders about the potential effect of contemporary global warming: sea-level rise; accelerated glacial retreat; changed atmospheric circulation patterns leading to changes in seasonal temperatures, increased storminess, and increased flooding?

France is a large country (two and a half times the size of the UK). There is an enormous diversity and range of French landscapes. This might affect where you plan to visit. British visitors might find the landscape in northern and western France not unlike British landscapes. Southern, Alpine and Mediterranean France may have more appeal. Those areas certainly have more dramatic geomorphology. Whatever! There is one resource you might find useful wherever you plan to visit. The French maintain a comprehensive system of Marine Parks, National Parks, and more local Natural Parks (Fig. 16.1). The purpose of the Parks is the conservation of landscape and wildlife. The parks are often designated and delimited on the basis of the land-scape (therefore on the basis of geomorphology). They are superbly managed, and local "Syndicats d'Initiative" usually have excellent informative guide-type material.

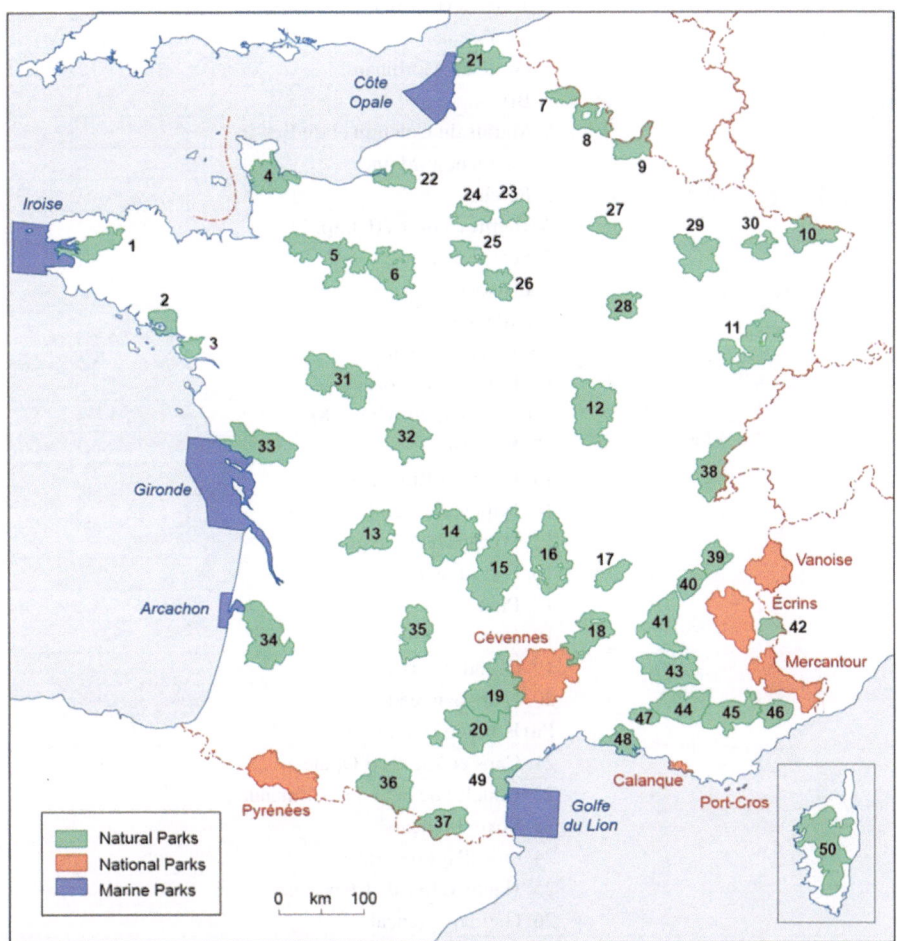

Fig. 16.1 Map of conservation areas in France see Table 16.1

Table 16.1 French Natural Parks and other conservation areas (located on Fig. 16.1)	**National Parks** (located on Fig. 16.1)
	Vanoise
	Écrins
	Mercantour
	Port-Cros
	Calanques
	Cévennes
	Pyrenees
	Regional (Natural) Parks (located by number on Fig. 16.1
	Grouped in relation to the preceding regional Chaps. (6, 7, 8, 9, 10, 11, 12, 13, 14, 15).
	Armorica (Chap. 6)
	1. Armorique
	2. Golfe du Morbihan
	3. Brière
	4. Marais du Cotentin et du Bessin
	5. Normandie-Maine
	6. Perche
	Ardennes/Vosges (Chap. 7)
	7. Scarpe-Escaut
	8. L'Avesnois
	9. Ardennes
	10. Vosges du Nord
	11. Ballons des Vosges
	Massif Central (Chap. 8)
	12. Morvan
	13. Périgord-Limousin
	14. Millevaches en Limousin
	15. Volcans d'Auvergne
	16. Livradois-Forez
	17. Pilat
	18. Monts d'Ardèche
	19. Grands Causses
	20. Haut-Languedoc
	Paris Basin (Chap. 9)
	21. Caps et Marais d'Opale
	22. Boucles de la Seine Normande
	23. Vexin Français
	24. Oise-Pays de France
	25. Haute Vallée de Chevreuse
	26. Gâtinais Français
	27. Montagne de Reims
	28. Fôret d'Orient
	29/30. Lorraine (two sections: West and east)
	31. Loire-Anjou-Touraine

(continued)

Table 16.1   (continued)

32. La Brenne
Aquitaine (Chap. 10)
33. Marais Poitevin
34. Landes de Gascogne
35. Causses du Quercy
Pyrenees (Chap. 11)
36. Pyrenees Ariègeoises
37. Pyrenees Catalanes
Jura (Chap. 12)
38. Haut-Jura
Alps (north) (Chap. 13)
39. Massif des Bauges
40. Chartreuse
41. Vercours
42. Queyras
Alps (south and coast) (Chap. 14)
43. Baronnies Provençales
44. Luberon
45. Verdon
46. Préalpes d'Azur
47. Les Alpilles
48 Camargue
49. Narbonnaise Méditerranée
Corsica (Chap. 15)
50. Corse
Marine Parks
d'Iroise
Estuaires Picards et la mer d'Opale
Estuaire de la Gironde et de la mer des Pertuis
Bassin d'Arcachon
Golfe du Lion

Logistics and Further Information

I assume that the readership of this book would primarily be British and be intending to visit France for educational, research or purely for recreational purposes. I make a few comments on possible reading below, but first we need to consider some of the logistics of travel. Up until the end of 2020, the UK had been part of the EU. British citizens had rights as European citizens to visit, travel, own property, and be covered for basic healthcare. All this has changed with BREXIT. To visit France we now need healthcare cover, we might need international driving licences, and it looks as though we might even need visas. In the past I have benefited from our EU membership, to some extent in France but more so through professional research links in Spain. I suspect that in future neither would be easy, perhaps not even possible. I didn't notice "citizens' rights" having much priority in the BREXIT negotiations! At this point, I'd better stop writing, or I might become guilty of "ranting".

A. Harvey, *The Geomorphology of French Landscapes*, https://doi.org/10.1007/978-3-031-68490-6

Background Reading

I recommend a few books as background reading (there is not a lot in English) Much of the most recent research literature is not appropriate anyway. I do however, give a rather short list of very specific references mentioned in the text (see below).

For general background in geology and geomorphology, though not specific to France, I recommend two introductory books, both published by Dunedin Press, Edinburgh: (Now under the University of Liverpool Press).

Introducing Geology, by Graham Park, third ed. 2018,

Introducing Geomorphology (by me), Adrian Harvey, second ed. 2022.

For a more advanced cover of the geological evolution of France within its European context, see another Dunedin book, also by Graham Park: The Making of Europe, A Geological History, 2014, Dunedin Press, Edinburgh; (Now under the University of Liverpool Press).

I would recommend a few books for general tourism (though not always good on the geomorphology!). Either the regional Michelin guides or the AA/Hachette Guide to France might be useful. An excellent book on French rural landscapes, dealing almost entirely with the human aspect rather than on the physical landscapes (or the geomorphology) is: L'Atlas des Paysages Ruraux de France, Edited by Pierre Brunet, and published by Jean-Pierre de Monza, Paris, 1992. For more specific local geology of the vineyard areas [useful local maps; local geological excursions described]. The Wines and Winelands of France: Geological Journeys, by Charles Pomerol, Robertson McCarta Publishers, London, 1989.

I recommend two modern books on the geomorphology of France. Where appropriate, in my Chaps. 6–15 I refer by author to specific chapters from these two books.

Monique Fort and Marie-Francoise André (Editors), Landscapes and Landforms of France, Springer, Dordrecht, 2014. Despite some shortcomings, this book does include some superb site-specific chapters on some critical locations in French geomorphology.

Denis Mercier (Direction), Geomorphologie de la France, Dunod publ. Paris, 2013. This is the one of the only works I refer to written in French. It is an excellent, comprehensive overview of the geomorphology of France.

References

Antoine P, Limondin-Louzouet N, Chaussé C, Latridou J-P, Pastre J-F, Auguste P, Bahain J-J, Falguères C, Ghaleb B (2007) Pleistocene fluvial terraces from northern France (Seine, Yonne, Somme): synthesis, and new results from interglacial deposits. Quat Sci Rev 26:2701–2723

Astrade L, Jacob-Rousseau N, Bravard J-P, Allignol F, Simac L (2011) Detailed chronology of mid-altitude fluvial system response to changing climate and societies at the end of the Little Ice Age (Southwestern Alps and Cevennes, France). Geomorphology 37:100–116

Bossard V, Lerma AN (2020) Geomorphologic characteristics and evolution of managed dunes on the South West Coast of France. Geomorphology 367:107312

Bravard J-P (1989) La métamorphose des rivières des Alpes fancaises â la fin du Moyen Âge et â l'Époque moderne. Bull Société Géographique de Liège 25:515–521

Buoncristiani J-F, Campy M (2011) Quaternary glaciations in the French Alps and Jura. In: Ehlers J, Gibbard PL, Hughes PD (eds) Developments in quaternary science. Elsevier, Amsterdam, pp 117–126

Calvet M, Gunnell Y (2008) Planar landforms as markers of denudation chronology: an inversion of East Pyrenean tectonics and sedimentary basin analysis, vol 296. Geological Society, London, Special Publication, pp 147–166

Conchon O (1978) Quaternary studies in Corsica (France). Quat Res 9(1):41–53

Conchon O (1986) Quaternary glaciations in Corsica. In: Sibana V, Bowen DQ, Richmond GM (eds) Quaternary glaciations in the Northern Hemisphere, vol 5. Quaternary Science Reviews, pp 429–432

Cordier S, Harmand D, Frenchen M, Beiner M (2006) Fluvial system response to Middle and Upper Pleistocene climatic change in the Meurthe and Moselle valleys (Eastern Paris Basin and Rhenish Massif). Quat Sci Rev 25:1460–1474

Cossart E, Braucher R, Fort M, Bourles DL, Carcaillet J (2008) Slope instability in relation to glacial debuttressing in Alpine areas (Upper Durance catchment, southeastern France): evidence from field data and 10Be cosmic ray exposure ages. Geomorphology 95:3–26

Delmas M (2005) La déglaciation dans le massif du Carlit (Pyrenees orientales): approches géomorphologique et géochronologique nouvelles. Quaternaire 16:45–55

Demoulin A, Hallot E (2009) Shape and amount of the Quaternary uplift of the Rheinish shield and the Ardennes (western Europe). Tectonophysics 474:696–708

Fort M-F, André F (Eds) (2014) Landscapes and Landforms of France, Springer, Dordrecht, pp 274

Gautier E, Grivel S (2006) Multi-scale analysis of island formation and development in the Middle Loire River, France. In: Rowan JS, Duck RW, Werrity A (eds) Sediment dynamics and

© The Author(s), under exclusive license to Springer Nature
Switzerland AG 2025
A. Harvey, *The Geomorphology of French Landscapes*,
https://doi.org/10.1007/978-3-031-68490-6

the hydro-morphology of fluvial systems, vol 306. International Association of Hydrologic Science, Publ, pp 179–187

Kuhlemann J, Frisch W, Szekwly B, Dunkl I, Danisk M, Krumei I (2005) Wurmian maximum glaciations in Corsica. Austrian Journal of Earth Sciences 97:68–81

Lespez L, Clet-Pellerin M, Limondin-Lozouet N, Pastre JF, Fontugne M, Marcigny CA (2008) Fluvial system evolution and environmental changes during the Holocene in the Mue valley (W France). Geomorphology 98:55–70

Liébault F, Clément P, Piégay H, Roger CF, Kondolf GM, Landon N (2002) Contemporary channel changes in the Eygues basin, southern French Prealps: the relationship of sub-basin variability to watershed characteristics. Geomorphology 45:53–66

Liébault F, Piégay H (2002) Causes of the 20th century channel narrowing in mountain and piedmont rivers of Southeastern France. Earth Surf Process Landf 27:425–444

Oldfield F, Berthier F (2001) The multi-proxy late-Pleistocene and Holocene record from the sediments of the Grand Lac d'Annecy, eastern France. J Palaeolimnol 25(2):133–135

Pastre JF, Limondin-Lozouet N, Leroyer C, Ponel P (2003) River system evolution and environmental changes during the Late Glacial in the Paris Basin (France). Quat Sci Rev 22:2177–2188

Regnauld H, Jennings S, Delaney C, Lemasson L (1996) Holocene sea level variations and geomorphological response: an example in Northern Brittany. Quat Sci Rev 15:781–787

Tourenq J, Pomerol C (1995) Mise en evidence par le presence d'augite du Massif Central de l'existence d'une pré-Loire, pré-Seine coulant vers la Manche au Pleistocene. Comptes Rendus Academie Scientifique, Paris 320:1163–1169

Vanara N, Nicod J (2013) Karsts de France. In: Mercier D (dir) Géomorphologie de France. Dunod, Paris, 37–48

Vayssière A et al (10 co-authors) (2020) Readjustments of a sinuous river during the last 6000 years in northwestern Europe (Cher River, France) from an active meandering river to a stable river course under human forcing. Geomorphology 370:107395

Vella C, Provansal M (2000) Relative sea-level rise and neotectonic events during the last 6500 yrs on the southern eastern Rhône delta, France. Mar Geol 170:27–39

Wooldridge SW, Linton DL (1955) Structure, surface and drainage in south-east England. Philip and Son, London, p 176